Computational Biology

T0143037

The *Computational Biology* series publishes the very latest, high-quality research devoted to specific issues in computer-assisted analysis of biological data. The main emphasis is on current scientific developments and innovative techniques in computational biology (bioinformatics), bringing to light methods from mathematics, statistics and computer science that directly address biological problems currently under investigation.

The series offers publications that present the state-of-the-art regarding the problems in question; show computational biology/bioinformatics methods at work; and finally discuss anticipated demands regarding developments in future methodology. Titles can range from focused monographs, to undergraduate and graduate textbooks, and professional text/reference works.

Author guidelines: springer.com > Authors > Author Guidelines

For other titles published in this series, go to
http://www.springer.com/series/5769

Kun-Mao Chao • Louxin Zhang

Sequence Comparison

Theory and Methods

 Springer

Kun-Mao Chao, BS, MS, PhD
Department of Computer Science and
    Information Engineering,
National Taiwan University,
Taiwan

Louxin Zhang, BSc, MSc, PhD
Department of Mathematics,
National University of Singapore,
Singapore

Computational Biology Series ISSN 1568-2684
ISBN: 978-1-84996-782-2   e-ISBN: 978-1-84800-320-0
DOI 10.1007/978-1-84800-320-0

British Library Cataloguing in Publication Data
A catalogue record for this book is available from the British Library

Printed on acid-free paper

Springer Science+Business Media
springer.com

Foreword

My first thought when I saw a preliminary version of this book was: Too bad there was nothing like this book when I really needed it.

Around 20 years ago, I decided it was time to change my research directions. After exploring a number of possibilities, I decided that the area of overlap between molecular biology and computer science (which later came to be called "bioinformatics") was my best bet for an exciting career. The next decision was to select a specific class of problems to work on, and the main criterion for me was that algorithmic methods would be the main key to success. I decided to work on sequence analysis. A book like this could have, so to speak, straightened my learning curve.

It is amazing to me that those two conclusions still apply: bioinformatics is a tremendously vibrant and rewarding field to be in, and sequence comparison is (arguably, at least) the subfield of bioinformatics where algorithmic techniques play the largest role in achieving success. The importance of sequence-analysis methods in bioinformatics can be measured objectively, simply by looking at the numbers of citations in the scientific literature for papers that describe successful developments; a high percentage of the most heavily cited scientific publications in the past 30 years are from this new field. Continued growth and importance of sequence analysis is guaranteed by the explosive development of new technologies for generating sequence data, where the cost has dropped 1000-fold in the past few years, and this fantastic decrease in cost means that sequencing and sequence analysis are taking over jobs that were previously handled another way.

Careful study of this book will be valuable for a wide range of readers, from students wanting to enter the field of bioinformatics, to experienced users of bioinformatic tools wanting to use tool options more intelligently, to bioinformatic specialists looking for the killer algorithm that will yield the next tool to sweep the field. I predict that you will need more that just mastery of this material to reach stardom in bioinformatics – there is also a huge amount of biology to be learned, together with a regular investment of time to keep up with the latest in data-generation technology and its applications. However, the material herein will remain useful for years, as new sequencing technologies and biological applications come and go.

I invite you to study this book carefully and apply ideas from it to one of the most exciting areas of science. And be grateful that two professionals with a combined 30 years of experience have taken the time to open the door for you.

State College, Pennsylvania   *Webb Miller*

June 2008

Preface

Biomolecular sequence comparison is the origin of bioinformatics. It has been extensively studied by biologists, computer scientists, and mathematicians for almost 40 years due to its numerous applications in biological sciences. Today, homology search is already a part of modern molecular biology. This book is a monograph on the state-of-the-art study of sequence alignment and homology search.

Sequence alignment, as a major topic of bioinformatics, is covered in most bioinformatics books. However, these books often tell one part of the story. The field is evolving. The BLAST program, a pearl of pearls, computes local alignments quickly and evaluates the statistical significance of any alignments that it finds. Although BLAST homology search is done more than 100,000 times per day, the statistical calculations used in this program are not widely understood by its users. In fact, these calculations keep on changing with advancement of alignment score statistics. Simply using BLAST without a reasonable understanding of its key ideas is not very different from using a PCR without knowing how PCR works. This is one of the motivations for us to write this book. It is intended for covering in depth a full spectrum of the field from alignment methods to the theory of scoring matrices and to alignment score statistics.

Sequence alignment deals with basic problems arising from processing DNA and protein sequence information. In the study of these problems, many powerful techniques have been invented. For instance, the filtration technique, powered with spaced seeds, is shown to be extremely efficient for comparing large genomes and for searching huge sequence databases. Local alignment score statistics have made homology search become a reliable method for annotating newly sequenced genomes. Without doubt, the ideas behind these outstanding techniques will enable new approaches in mining and processing structural information in biology. This is another motivation for us to write this book.

This book is composed of eight chapters and three appendixes. Chapter 1 works as a tutorial to help all levels of readers understand the connection among the other chapters. It discusses informally why biomolecular sequences are compared through alignment and how sequence alignment is done efficiently.

Chapters 2 to 5 form the method part. This part covers the basic algorithms and methods for sequence alignment. Chapter 2 introduces basic algorithmic techniques that are often used for solving various problems in sequence comparison.

In Chapter 3, we present the Needleman-Wunsch and Smith-Waterman algorithms, which, respectively, align a pair of sequences globally and locally, and their variants for coping with various gap penalty costs. For analysis of long genomic sequences, the space restriction is more critical than the time constraint. We therefore introduce an efficient space-saving strategy for sequence alignment. Finally, we discuss a few advanced topics of sequence alignment.

Chapter 4 introduces four popular homology search programs: FASTA, BLAST family, BLAT, and PatternHunter. We also discuss how to implement the filtration idea used in these programs with efficient data structures such as hash tables, suffix trees, and suffix arrays.

Chapter 5 covers briefly multiple sequence alignment. We discuss how a multiple sequence alignment is scored, and then show why the exact method based on a dynamic-programming approach is not feasible. Finally, we introduce the progressive alignment approach, which is adopted by ClustalW, MUSCLE, YAMA, and other popular programs for multiple sequence alignment.

Chapters 6 to 8 form the theory part. Chapter 6 covers the theoretic aspects of the seeding technique. PatternHunter demonstrates that an optimized spaced seed improves sensitivity substantially. Accordingly, elucidating the mechanism that confers power to spaced seeds and identifying good spaced seeds become new issues in homology search. This chapter presents a framework of studying these two issues by relating them to the probability of a spaced seed hitting a random alignment. We address why spaced seeds improve homology search sensitivity and discuss how to design good spaced seeds.

The Karlin-Altschul statistics of optimal local alignment scores are covered in Chapter 7. Optimal segment scores are shown to follow an extreme value distribution in asymptotic limit. The Karlin-Altschul sum statistic is also introduced. In the case of gapped local alignment, we describe how the statistical parameters of the distribution of the optimal alignment scores are estimated through empirical approach and discuss the edge-effect and multiple testing issues. We also relate theory to the calculations of the Expect and P-values in BLAST program.

Chapter 8 is about the substitution matrices. We start with the reconstruction of popular PAM and BLOSUM matrices. We then present Altschul's theoretic-information approach to scoring matrix selection and recent work on compositional adjustment of scoring matrices for aligning sequences with biased letter frequencies. Finally, we discuss gap penalty costs.

This text is targeted to a reader with a general scientific background. Little or no prior knowledge of biology, algorithms, and probability is expected or assumed. The basic notions from molecular biology that are useful for understanding the topics covered in this text are outlined in Appendix A. Appendix B provides a brief introduction to probability theory. Appendix C lists popular software packages for pairwise alignment, homology search, and multiple alignment.

This book is a general and rigorous text on the algorithmic techniques and mathematical foundations of sequence alignment and homology search. But, it is by no means comprehensive. It is impossible to give a complete introduction to this field because it is evolving too quickly. Accordingly, each chapter concludes with the bibliographic notes that report related work and recent progress. The reader may ultimately turn to the research articles published in scientific journals for more information and new progress.

Most of the text is written at a level that is suitable for undergraduates. It is based on lectures given to the students in the courses in bioinformatics and mathematical genomics at the National University of Singapore and the National Taiwan University each year during 2002 – 2008. These courses were offered to students from biology, computer science, electrical engineering, statistics, and mathematics majors. Here, we thank our students in the courses we have taught for their comments on the material, which are often incorporated into this text.

Despite our best efforts, this book may contain errors. It is our responsibility to correct any errors and omissions. A list of errata will be compiled and made available at http://www.math.nus.edu.sg/~matzlx/sequencebook.

Taiwan & Singapore   *Kun-Mao Chao*
June 2008   *Louxin Zhang*

Acknowledgments

We are extremely grateful to our mentor Webb Miller for kindly writing the fore-word for this book. The first author particularly wants to thank Webb for introducing him to the emerging field of computational molecular biology and guiding him from the basics nearly two decades ago.

The second author is particularly thankful to Ming Li for guiding and encouraging him since his student time in Waterloo. He also thanks his collaborators Kwok Pui Choi, Aaron Darling, Minmei Hou, Yong Kong, Jian Ma, Bin Ma, and Franco Preparata, with whom he worked on the topics covered in this book. In addition, he would like to thank Kal Yen Kaow Ng and Jialiang Yang for reading sections of the text and catching some nasty bugs.

We also thank the following people for their inspiring conversations, suggestions, and pointers: Stephen F. Altschul, Vineet Bafna, Louis H.Y. Chen, Ross C. Hardison, Xiaoqiu Huang, Tao Jiang, Jim Kent, Pavel Pevzner, David Sankoff, Scott Schwartz, Nikola Stojanovic, Lusheng Wang, Von Bing Yap, and Zheng Zhang.

Finally, it has been a pleasure to work with Springer in the development of this book. We especially thank our editor Wayne Wheeler and Catherine Brett for patiently shepherding this project and constantly reminding us of the deadline, which eventually made us survive. We also thank the copy editor C. Curioli for valuable comments and the production editor Frank Ganz for assistance with formatting.

About the Authors

Kun-Mao Chao was born in Tou-Liu, Taiwan, in 1963. He received the B.S. and M.S. degrees in computer engineering from National Chiao-Tung University, Taiwan, in 1985 and 1987, respectively, and the Ph.D. degree in computer science from The Pennsylvania State University, University Park, in 1993. He is currently a professor of bioinformatics at National Taiwan University, Taipei, Taiwan. From 1987 to 1989, he served in the ROC Air Force Headquarters as a system engineer. From 1993 to 1994, he worked as a postdoctoral fellow at Penn State's Center for Computational Biology. In 1994, he was a visiting research scientist at the National Center for Biotechnology Information, National Institutes of Health, Bethesda, Maryland. Before joining the faculty of National Taiwan University, he taught in the Department of Computer Science and Information Management, Providence University, Taichung, Taiwan, from 1994 to 1999, and the Department of Life Science, National Yang-Ming University, Taipei, Taiwan, from 1999 to 2002. He was a teaching award recipient of both Providence University and National Taiwan University. His current research interests include algorithms and bioinformatics. He is a member of Phi Tau Phi and Phi Kappa Phi.

Louxin Zhang studied mathematics at Lanzhou University, earning his B.S. and M.S. degrees, and studied computer science at the University of Waterloo, where he received his Ph.D. He has been a researcher and teacher in bioinformatics and computational biology at National University of Singapore (NUS) since 1996. His current research interests include genomic sequence analysis and phylogenetic analysis. His research interests also include applied combinatorics, algorithms, and theoretical computer science. In 1997, he received a Lee Kuan Yew Postdoctoral Research Fellowship to further his research. Currently, he is an associate professor of computational biology at NUS.

Contents

Chapter 1
Introduction

1.1 Biological Motivations

A vast diversity of organisms exist on Earth, but they are amazingly similar. Every organism is made up of cells and depends on two types of molecules: DNA and proteins. DNA in a cell transmits itself from generation to generation and holds genetic instructions that guide the synthesis of proteins and other molecules. Proteins perform biochemical reactions, pass signals among cells, and form the body's components. DNA is a sequence of 4 nucleotides, whereas proteins are made of 20 amino acids arranged in a linear chain.

Due to the linear structure of DNA and protein, sequence comparison has been one of the common practices in molecular biology since the first protein sequence was read in the 1950s. There are dozens of reasons for this. Among these reasons are the following: sequence comparisons allow identification of genes and other conserved sequence patterns; they can be used to establish functional, structural, and evolutionary relationship among proteins; and they provide a reliable method for inferring the biological functions of newly sequenced genes.

A striking fact revealed by the abundance of biological sequence data is that a large proportion of genes in an organism have significant similarity with ones in other organisms that diverged hundreds of millions of years ago. Two mammals have as many as 99% genes in common. Humans and fruit flies even have at least 50% genes in common.

On the other hand, different organisms do differ to some degree. This strongly suggests that the genomes of existing species are shaped by evolution. Accordingly, comparing related genomes provides the best hope for understanding the language of DNA and for unraveling the evolutionary relationship among different species.

Two sequences from different genomes are homologous if they evolved from a common ancestor. We are interested in homologies because they usually have conserved structures and functions. Because species diverged at different time points, homologous proteins are expected to be more similar in closely related organisms than in remotely related ones.

When a new protein is found, scientists usually have little idea about its function. Direct experimentation on a protein of interest is often costly and time-consuming. As a result, one common approach to inferring the protein's function is to find by similarity search its homologies that have been studied and are stored in database. One remarkable finding made through sequence comparison is about how cancer is caused. In 1983, a paper appearing in *Science* reported a 28-amino-acid sequence for platelet derived growth factors, a normal protein whose function is to stimulate cell growth. By searching against a small protein database created by himself, Doolittle found that the sequence is almost identical to a sequence of υ-sis, an oncogene causing cancer in woolly monkeys. This finding changed the way oncogenesis had been seen and understood. Today, it is generally accepted that cancer might be caused by a normal growth gene if it is switched on at the wrong time.

Today, powerful sequence comparison methods, together with comprehensive biological databases, have changed the practice of molecular biology and genomics. In the words of Gilbert, Nobel prize winner and co-inventor of practical DNA sequencing technology,

> The new paradigm now emerging, is that all the 'genes' will be known (in the sense of being resident in databases available electronically), and that the starting point of biological investigation will be theoretical. (Gilbert, 1991, [76])

1.2 Alignment: A Model for Sequence Comparison

1.2.1 Definition

Alignment is a model used by biologists for bringing up sequence similarity. Let Σ be an alphabet, and let $X = x_1 x_2 \ldots x_m$ and $Y = y_1 y_2 \ldots y_n$ be two sequences over Σ. We extend Σ by adding a special symbol '-', which denotes *space*. An *alignment* of sequences X and Y is a two-row matrix A with entries in $\Sigma \cup \{-\}$ such that

1. The first (second) row contains the letters of X (Y) in order;
2. One or more spaces '-' may appear between two consecutive letters of Σ in each row;
3. Each column contains at least one letter of Σ.

For example, an alignment between a fragment of the yeast glutathione S-transferase I (GST1) protein sequence and the viral-enhancing factor (VEF) protein sequence is shown in Figure 1.1.

In an alignment, columns that contain two letters are called *matches* if the letters are identical and *mismatches* otherwise. Columns containing a space are called *indels*. In order to be accurate, we sometimes call the columns *insertions* if they contain a space in the first row and *deletions* if they contain a space in the second row. In the alignment of sequences GST1 and VEF given in Figure 1.1, there are 37 matches highlighted by using vertical bars in the central line. Besides, there are 3 insertions and 15 deletions.

```
GST1  SRAFRLLWLLDHLNLEYEIVPYKRDANFRAPPELKKIHPLGRSPLLEVQDRETGKKKILA
       |  ||   | |   ||  |||    |  |          |       |  |  | |
VEF   SYLFRFRGLGDFMLLELQIVPILNLASVRVGNHHNGPHSYFNTTYLSVEVRDT-------

GST1  ESGFIFQYVL---QHFDHSHVLMSEDADIADQINYYLFYVEGSLQPPLMIEFILSKVKDS
       |  | |        | |               || |     |      | |  |
VEF   SGGVVFSYSRLGNEPMTHEH----HKFEVFKDYTIHLFIQE----PGQRLQLIVNKTLDT

GST1  GMPFPISYLARKVADKISQAYSSGEVKNQFDFV
       |        || |        |          |||
VEF   ALPNSQNIYARLTATQLVVGEQSIIISDDNDFV
```

Fig. 1.1 An example of protein sequence alignment.

From a biological point of view, an alignment of two sequences poses a hypothesis on how the sequences evolved from their closest common ancestor. An alignment accounts for the following three mutational events:

• Substitution (also called point mutation) - A nucleotide is replaced by another.
• Insertion - One or several nucleotides are inserted at a position in a sequence.
• Deletion - One of several nucleotides are deleted from a sequence.

In an alignment, a mismatch represents a hypothetical substitution event, and an indel $\binom{-}{a}$ or $\binom{a}{-}$ represents that a is added or deleted in an insertion or deletion event.

We often say an alignment is *gapped* or *ungapped* to emphasize whether indels are allowed or not in the alignment.

1.2.2 Alignment Graph

There are many ways of aligning two sequences. By using theoretic-graph concepts, however, we can obtain a compact representation of these alignments.

Consider two sequences of length m and n. The *alignment graph* of these two sequences is a direct graph G. The vertices of G are lattice points (i, j), $0 \le i \le m$ and $0 \le j \le n$ in the $(m+1) \times (n+1)$ grid, and there is an edge from vertex (i, j) to another (i', j') if and only if $0 \le i' - i \le 1$ and $0 \le j' - j \le 1$. Figure 1.2 gives the alignment graph for sequences ATACTGG and GTCCGTG. As shown in Figure 1.2, there are vertical, horizontal, and diagonal edges. In a directed graph, a sink is a vertex that has every incident edge directed toward it, and a source is a vertex that has every incident edge directed outwards. In the alignment graph, $(0,0)$ is the unique source and (m,n) is the unique sink.

The alignment graph is very useful for understanding and developing alignment algorithms as we shall see in Chapter 3. One reason for this is that all possible alignments correspond one-to-one to the directed paths from the source $(0,0)$ to the sink (m,n). Hence, it gives a compact representation of all possible alignments of

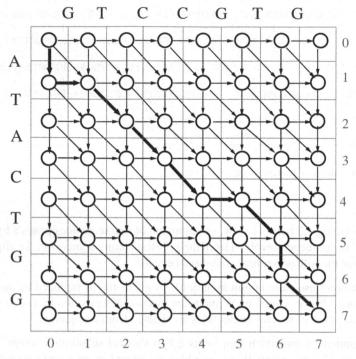

Fig. 1.2 Alignment graph for sequences ATACTGG and GTCCGTG.

the sequences. To see the correspondence, we consider the alignment graph given in Figure 1.2 in what follows.

Let $X = $ ATACTGG be the first sequence and $Y = $ GTCCGTG be the second sequence. The highlighted path in the alignment graph is

$$s \rightarrow (1,0) \rightarrow (1,1) \rightarrow (2,2) \rightarrow (3,3) \rightarrow (4,4) \rightarrow (4,5) \rightarrow (5,6) \rightarrow (6,6) \rightarrow (7,7),$$

where s is the source $(0,0)$. Let x_i (y_i) denote the ith letter of the first (second) sequence. Walking along the path, we write down a column $\binom{x_i}{y_j}$, $\binom{x_i}{-}$, and $\binom{-}{y_j}$ if we see a diagonal, vertical, and horizontal edge entering vertex (i, j), respectively. After reaching the sink $(7, 7)$, we obtain the following nine-column alignment of X and Y:

$$\begin{array}{ll} X & \text{A} - \text{T A C} - \text{T G G} \\ Y & - \text{G T C C G T} - \text{G} \end{array}$$

where the ith column corresponds to the ith edge in the path.

On the other hand, given a k-column alignment of X and Y, we let u_i (and v_i) be the number of letters in the first i columns in the row of X (and Y) for $i = 0, 1, \ldots, k$. Trivially, $u_0 = v_0 = 0$ and $u_k = m$ and $v_k = n$. For any i, we also have that

$$0 \leq u_{i+1} - u_i \leq 1$$

and

$$0 \leq v_{i+1} - v_i \leq 1.$$

This fact implies that there is an edge from (u_i, v_i) to (u_{i+1}, v_{i+1}). Thus,

$$(u_0, v_0) \rightarrow (u_1, v_1) \rightarrow \ldots \rightarrow (u_k, v_k)$$

is a path from the source to the sink. For example, from alignment

$$X \quad A\ T\ A\ C\ T\ G\ -\ G$$
$$Y \quad G\ T\ C\ C\ -\ G\ T\ G,$$

we obtain the following path

$$(0,0) \rightarrow (1,1) \rightarrow (2,2) \rightarrow (3,3) \rightarrow (4,4) \rightarrow (5,4) \rightarrow (6,5) \rightarrow (6,6) \rightarrow (7,7).$$

In summary, an alignment of two sequences corresponds uniquely to a path from the source to the sink in the alignment graph of these two sequences. In this correspondence, match/mismatches, deletions, and insertions correspond to diagonal, vertical, and horizontal edges, respectively.

The aim of alignment is to find the best one of all the possible alignments of two sequences. Hence, a natural question to ask is: How many possible alignments are there for two sequences of length m and n? This is equivalent to asking how many different paths there are from the source node $(0,0)$ to the sink node (m,n) in an alignment graph. For a simple problem of aligning $x_1 x_2$ and y_1, all 5 possible alignments are shown below:

$$
\begin{array}{ccccc}
x_1\ x_2 & x_1\ x_2 & -\ x_1\ x_2 & x_1\ x_2\ - & x_1\ -\ x_2 \\
y_1\ - & -\ y_1 & y_1\ -\ - & -\ -\ y_1 & -\ y_1\ -
\end{array}
$$

We denote the number of possible alignments by $a(m,n)$. In an alignment graph, the paths from $(0,0)$ to a vertex (i,j) correspond to possible alignments of two sequences of length i and j. Hence, $a(i,j)$ is equal to the number of paths from $(0,0)$ to the vertex (i,j). Because every path from the source $(0,0)$ to the sink (m,n) must go through exactly one of the vertices $(m-1,n)$, $(m,n-1)$, and $(n-1,m-1)$, we obtain

$$a(m,n) = a(m-1,n) + a(m,n-1) + a(n-1,m-1). \tag{1.1}$$

Noting that there is only one way to align the empty sequence to a non-empty sequence, we also have that

$$a(0,k) = a(k,0) = 1 \tag{1.2}$$

for any $k \geq 0$.

Table 1.1 The number of possible alignments of two sequences of length m and n for $n, m \le 8$.

	0	1	2	3	4	5	6	7	8
0	1	1	1	1	1	1	1	1	1
1		3	5	7	9	11	13	15	17
2			13	25	41	61	85	113	145
3				63	129	231	377	575	833
4					321	681	1,289	2,241	3,649
5						1,683	3,653	7,183	13,073
6							8,989	19,825	40,081
7								48,639	108,545
8									265,729

Recurrence relation (1.1) and basis conditions (1.2) enable us to calculate $a(m, n)$ efficiently. Table 1.1 gives the values of $a(m, n)$ for any $m, n \le 8$. As we see from the table, the number $a(m, n)$ grows quite fast with m and n. For $n = m = 8$, this number is already 265,729! As the matter of fact, we have that (see the gray box below)

$$a(m, n) = \sum_{k=\max\{n,m\}}^{m+n} \binom{k}{m}\binom{m}{n+m-k}. \tag{1.3}$$

In particular, when $m = n$, the sum of the last two terms of the right side of (1.3) is

$$\binom{2n-1}{n}\binom{n}{1} + \binom{2n}{n} = \frac{(n+2)(2n)!}{2(n!)^2}.$$

Applying Stirling's approximation

$$x! \approx \sqrt{2\pi}x^{x+1/2}e^{-x},$$

we obtain $a(100, 100) > 2^{200} \approx 10^{60}$, an astronomically large number! Clearly, this shows that it is definitely not feasible to examine all possible alignments and that we need an efficient method for sequence alignment.

Proof of formula (1.3)

We first consider the arrangements of letters in the first row and then those in the second row of a k-column alignment. Assume there are m letters in the first row. There are $\binom{k}{m}$ ways to arrange the letters in the first row. Fix such an arbitrary arrangement. By the definition of alignment, there are $k - n$ spaces in the second row, and these spaces must be placed below the letters in the first row. Thus, there are $\binom{m}{k-n} = \binom{m}{m+n-k}$ arrangements for the second row. By the multiplication principle, there are

$$a_k(m,n) = \binom{k}{m}\binom{m}{m+n-k}$$

possible alignments of k columns.

Each alignment of the sequences has at least $\max\{m,n\}$ and at most $m+n$ columns. Summing $a_k(m,n)$ over all k from $\max\{m,n\}$ to $m+n$ yields formula (1.3).

1.3 Scoring Alignment

Our goal is to find, given two DNA or protein sequences, the best alignment of them. For this purpose, we need a rule to tell us the goodness of each possible alignment. The earliest similarity measure was based on percent identical residues, that is, simply to count matches in an alignment. In the old days, this simple rule worked because it was rare to see the low percent identity of two proteins with similar functions like homeodomain proteins. Nowadays, a more general rule has to be used to score an alignment.

First, we have a score $s(a,b)$ for matching a with b for any pair of letters a and b. Usually, $s(a,a) > 0$, and $s(a,b) < 0$ for $a \neq b$. Assume there are k letters $a_1, a_2, \ldots, a_k$. All these possible scores are specified by a matrix $(s(a_i, b_j))$, called a *scoring matrix*. For example, when DNA sequences are aligned, the following scoring matrix may be used:

	A	G	C	T
A	2	-1	-1	-1
G	-1	2	-1	-1
C	-1	-1	2	-1
T	-1	-1	-1	2

This scoring matrix indicates that all matches score 2 whereas mismatches are penalized by 1. Assume we align two sequences. If one sequence has letter a and another has b in a position, it is unknown whether a had been replaced by b or the other way around in evolutionary history. Thus, scoring matrices are usually symmetric like the one given above. In this book, we only write down the lower triangular part of a scoring matrix if it is symmetric.

Scoring matrices for DNA sequence alignment are usually simple. All different matches score the same, and so do all mismatches. For protein sequences, however, scoring matrices are quite complicated. Frequently used scoring matrices are developed using statistical analysis of real sequence data. They reflect the frequency of an amino acid replacing another in biologically related protein sequences. As a result, a scoring matrix for protein sequence alignment is usually called a *substitu-*

tion matrix. We will discuss in detail how substitution matrices are constructed and selected for sequence alignment in Chapter 8.

Second, we consider indels. In an alignment, indels and mismatches are introduced to bring up matches that appear later. Thus, indels are penalized like mismatches. The most straightforward method is to penalize each indel by some constant δ. However, two or more nucleotides are frequently inserted or deleted together as a result of biochemical processes such as replication slippage. Hence, penalizing a gap of length k by $-k\delta$ is too cruel. A *gap* in an alignment is defined as a sequence of spaces locating between two letters in one row. A popular gap penalty model, called *affine gap penalty*, scores a gap of length k as

$$-(o+k \times e),$$

where $o > 0$ is considered as the penalty for opening a gap and $e > 0$ is the penalty for extending a gap by one letter. The opening gap penalty o is usually big whereas the gap extension penalty e is small. Note that simply multiples of the number of indels is a special case of the affine gap penalty model in which $o = 0$.

A scoring matrix and a gap penalty model form a *scoring scheme* or a *scoring system*. With a scoring scheme in hand, the score of an alignment is calculated as the sum of individual scores, one for each aligned pair of letters, and scores for gaps. Consider the comparison of two DNA sequences with the simple scoring matrix given above, which assigns 2 to each match and -1 to each mismatch. If we simply penalize each indel by -1.5, the score for the alignment on page 4 is

$$-1.5 - 1.5 + 2 - 1 + 2 - 1.5 + 2 - 1.5 + 2 = 3.$$

As we will see in Section 8.3, in any scoring matrix, the substitution score $s(a,b)$ is essentially a logarithm of the ratio of the probability that we expect to see a and b aligned in biologically related sequences to the probability that they are aligned in unrelated random sequences. Hence, being the sum of individual log-odds scores, the score of a ungapped alignment reflects the likelihood that this alignment was generated as a consequence of sequence evolution.

1.4 Computing Sequence Alignment

In this section, we briefly define the global and local alignment problems and then relate the alignment problem to some interesting algorithmic problems in computer science, mathematics, and information theory.

1.4.1 Global Alignment Problem

With a scoring system, we associate a score to each possible alignment. The *optimal alignments* are those with the maximum score. The *global alignment problem* (or simply alignment problem) is stated as

Global Alignment Problem

Input: Two sequences x and y and a scoring scheme.
Solution: An optimal alignment of x and y (as defined by the scoring scheme).

Because there is a huge number of possible alignments for two sequences, it is not feasible to find the optimal one by examining all alignments one by one. Fortunately, there is a very efficient algorithm for this problem. This algorithm is now called the Needleman-Wunsch algorithm. The so-called dynamic programming idea behind this algorithm is so simple that such an algorithm has been discovered and rediscovered in different form many times. The Needleman-Wunsch algorithm and its generalizations are extensively discussed in Chapter 3.

The sequence alignment problem seems quite simple. But, it is rather general as being closely related to several interesting problems in mathematics, computer science, and information theory. Here we just name two such examples. The *longest common subsequence problem* is, given two sequences, to find a longest sequence whose letters appear in each sequence in order, but not necessarily in consecutive positions. This problem had been interested in mathematics long before Needleman and Wunsch discovered their algorithm for aligning DNA sequences. Consider a special scoring system $\mathscr{S}$ that assigns 1 to each match, $-\infty$ to each mismatch, and 0 to each indel. It is easy to verify that the optimal alignment of two sequences found using $\mathscr{S}$ must not contain mismatches. As a result, all the matches in the alignment give a longest common subsequence. Hence, the longest common subsequence problem is identical to the sequence alignment problem under the particular scoring system $\mathscr{S}$.

There are two ways of measuring the similarity of two sequences: similarity scores and distance scores. In distance scores, the smaller the score, the more closely related are the two sequences. Hamming distance allows one only to compare sequences of the same length. In 1966, Levenshtein introduced *edit distance* for comparison of sequences of different lengths. It is defined as the minimum number of editing operations that are needed to transform one sequence into another, where the editing operations include insertion of a letter, deletion of a letter, and substitution of a letter for another. It is left to the reader to find out that calculating the edit distance between two strings is equivalent to the sequence alignment problem under a particular scoring system.

1.4.2 Local Alignment Problem

Proteins often have multiple functions. Two proteins that have a common function
may be similar only in functional domain regions. For example, homeodomain proteins, which play important roles in developmental processes, are present in a variety of species. These proteins in different species are only similar in one domain of
about 60 amino acids long, encoded by homeobox genes. Obviously, aligning the
entire sequences will not be useful for identification of the similarity among homeodomain proteins. This raises the problem of finding, given two sequences, which
respective segments have the best alignments. Such an alignment between some
segments of each sequence is called *local alignment* of the given sequences. The
problem of aligning locally sequences is formally stated as

Local Alignment Problem

Input: Two sequences $x = x_1 x_2 \ldots x_m$ and $y = y_1 y_2 \ldots y_n$ and a scoring
scheme.
Solution: An alignment of fragments $x_i x_{i+1} \ldots x_j$ and $y_k y_{k+1} \ldots y_l$, that has
the largest score among all alignments of all pairs of fragments of x and y.

A straightforward method for this problem is to find the optimal alignment for
every pair of fragments of x and y using the Needleman-Wunsch algorithm. The se-
quence x has $\binom{m}{2}$ fragments and y has $\binom{n}{2}$ ones. Thus, this method is rather inefficient
because its running time will increase by roughly $m^2 n^2$ times. Instead, applying directly the dynamic programming idea leads to an algorithm that is as efficient as
the Needleman-Wunsch algorithm although it is a bit more tricky this time. This
dynamic programming algorithm, called *Smith-Waterman* algorithm, is covered in
Section 3.4.

Homology search is one important application of the local alignment problem. In
this case, we have a query sequence, say, a newly sequenced gene, and a database.
We wish to search the entire database to find those sequences that match locally (to
a significant degree) with the query. Because databases have easily millions of sequences, Smith-Waterman algorithm, having quadratic-time complexity, is too demanding in computational time for homology search. Accordingly, fast heuristic
search tools have been developed in the past two decades. Chapter 4 will present
several frequently used homology search tools.

Filtration is a useful idea for designing fast homology search programs. A
filtration-based program first identifies short exact matches specified by a fixed pattern (called seed) of two sequences and then extends each match to both sides for
local alignment. A clever technique in speeding up homology search process is to
substitute optimized spaced seed for consecutive seed as exemplified in Pattern-
Hunter. Theoretic treatment of spaced seed technique is studied in Chapter 6.

1.5 Multiple Alignment

An alignment of two sequences is called an *pairwise* alignment. The above definition of pairwise alignment can be straightforwardly generalized to the case of multiple sequences. Formally, a *multiple alignment* of k sequences $X_1, X_2, \ldots, X_k$ over an alphabet Σ is specified by a $k \times n$ matrix M. Each entry of M is a letter of Σ or a space '-', and each row j contains the letters of sequence X_j in order, which may be interspersed with '-'s. We request that each column of the matrix contains at least one letter of Σ. Below is a multiple alignment of partial sequences of five globin proteins:

```
Hb_a      LSPADKTNVUAAWGKVGA----HAGEYGAE
Hb_b      LTPEEKSAVTALWGKV------NVDEVGGE
Mb_SW     LSEGEWQLVLHVWAKVEA----DVAGHGQD
LebHB     LTESQAALVKSSWEEFNA----NIPKHTHR
BacHB     QTINIIKATVPVLKEHG------V-TITTT
```

Multiple alignment is often used to assess sequence conservation of three or more closely related proteins. Biologically similar proteins may have very diverged sequences and hence may not exhibit a strong sequence similarity. Comparing many sequences at the same time often finds weak similarities that are invisible in pairwise sequence comparison.

Several issues arise in aligning multiple sequences. First, it is not obvious how to score a multiple alignment. Intuitively, high scores should correspond to highly conserved sequences. One popular scoring method is the *Sum-of-Pairs (SP) score*. Any alignment A of k sequences $x_1, x_2, \ldots, x_k$ gives a pairwise alignment $A(x_i, x_j)$ of x_i and x_j when restricted to these two sequences. We use $s(i, j)$ to denote the score of $A(x_i, x_j)$. The SP score of the alignment A is defined as

$$SP(A) = \sum_{1 \le i < j \le k} s(i, j).$$

Note that the SP score is identical to the score of a pairwise alignment when there are only two sequences. The details of the SP score and other scoring methods can be found in Section 5.2

Second, aligning multiple sequences is extremely time-consuming. The SP score of a multiple alignment is a generalization of a pairwise alignment score. Similarly, the dynamic programming algorithm can generalize to multiple sequences in a straightforward manner. However, such an algorithm will use roughly $(2m)^k$ arithmetic operations for aligning k sequences of length m. For small k, it works well. The running time is simply too much when k is large, say 30. Several heuristic approaches have been proposed for speeding up multiple alignment process (see Section 5.4).

1.6 What Alignments Are Meaningful?

Although homology and similarity are often interchanged in popular usage, they are completely different. Homology is qualitative, which means having a common ancestor. On the other hand, similarity refers to the degree of the match between two sequences. Similarity is an expected consequence of homology, but not a necessary one. It may occur due to chance or due to an evolutionary process whereby organisms independently evolve similar traits such as the wings of insect and bats.

Assume we find a good match for a newly sequenced gene through database search. Does this match reflect a homology? Nobody knows what really happened over evolutionary time. When we say that a sequence is homologous to another, we are stating what we believe. No matter how high is the alignment score, we can never be 100% sure. Hence, a central question in sequence comparison is how frequently an alignment score is expected to occur by chance. This question has been extensively investigated through the study of the alignments of random sequences. The Karlin-Altschul alignment statistics covered in Chapter 7 lay the foundation for answering this important question.

To approach theoretically the question, we need to model biological sequences. The simplest model for random sequences assumes that the letters in all positions are generated independently, with probability distribution

$$p_1, p_2, \ldots, p_r$$

for all letters $a_1, a_2, \ldots a_r$ in the alphabet, where r is 4 for DNA sequences and 20 for protein sequences. We call it the *Bernoulli sequence model*.

The theoretical studies covered in Chapters 6 and 7 are based on this simple model. However, most of the results generalize to the high-order Markov chain model in a straightforward manner. In the *kth-order Markov chain sequence model*, the probability that a letter is present at any position j depends on the letters in the preceding k sites: $i - k, i - k + 1, \ldots, j + 1$. The third-order Markov chain model is often used to model gene coding sequences.

1.7 Overview of the Book

This book is structured into two parts. The first part examines alignment algorithms and techniques and is composed of four chapters, and the second part focuses on the theoretical issues of sequence comparison and has three chapters. The individual chapters cover topics as follows.

2 Basic algorithmic techniques. Starting with basic definitions and notions, we introduce the greedy, divide-and-conquer, and dynamic programming approaches that are frequently used in designing algorithms in bioinformatics.

3 Pairwise sequence alignment. We start with the dot matrix representation of pairwise alignment. We introduce the Needleman-Wunsch and Smith-Waterman algorithms. We further describe several variants of these two classic algorithms for coping with special cases of scoring schemes, as well as space-saving strategies for aligning long sequences. We also cover constrained sequence alignment and suboptimal alignment.

4 Homology search tools. After showing how filtration technique speeds up homology search process, we describe in detail four frequently used homology search tools: FASTA, BLAST, BLAT, and PatternHunter.

5 Multiple sequence alignment. Multiple sequence alignment finds applications in prediction of protein functions and phylogenetic studies. After introducing the sum-of-pairs score, we generalize the dynamic programming idea to aligning multiple sequences and describe how progressive approach speeds up the multiple alignment process.

6 Anatomy of spaced seeds. We focus on the theoretic analysis of spaced seed technique. Starting with a brief introduction to the spaced seed technique, we first discuss the trade-off between sensitivity and specificity of seeding-based methods for homology search. We then present a framework for the analysis of the hit probability of spaced seeds and address seed selection issues.

7 Local alignment statistics. We focus on the Karlin-Altschul statistics of local alignment scores. We show that optimal segment scores are accurately described by an extreme value distribution in asymptotic limit, and introduce the Karlin-Altschul sum statistic. In the case of gapped local alignment, we describe how the statistical parameters for the score distribution are estimated through empirical approach, and discuss the edge-effect and multiple testing issues. Finally, we illustrate how the Expect value and P-value are calculated in BLAST using two BLAST printouts.

8 Scoring matrices. We start with the frequently used PAM and BLOSUM matrices. We show that scoring matrices for aligning protein sequences take essentially a log-odds form and there is one-to-one correspondence between so-called valid scoring matrices and the sets of target and background frequencies. We also discuss how scoring matrices are selected and adjusted for comparing sequences of biased letter composition. Finally, we discuss gap score schemes.

1.8 Bibliographic Notes and Further Reading

After nearly 50 years of research, there are hundreds of available tools and thousands of research papers in sequence alignment. We will not attempt to cover all (or even a large portion) of this research in this text. Rather, we will be content to provide pointers to some of the most relevant and useful references on the topics not covered in this text.

For the earlier phase of sequence comparison, we refer the reader to the paper of Sankoff [174]. For information on probabilistic and statistical approach to sequence alignment, we refer the reader to the books by Durbin et al. [61], Ewens and Grant [64], and Waterman [197]. For information on sequence comparisons in DNA sequence assembly, we refer the reader to the survey paper of Myers [149] and the books of Gusfield [85] and Deonier, Tavaré and Waterman [58]. For information on future directions and challenging problems in comparison of genomic DNA sequences, we refer the reader to the review papers by Batzoglou [23] and Miller [138].

PART I. ALGORITHMS AND TECHNIQUES

Chapter 2
Basic Algorithmic Techniques

An *algorithm* is a step-by-step procedure for solving a problem by a computer. Although the act of designing an algorithm is considered as an art and can never be automated, its general strategies are learnable. Here we introduce a few frameworks of computer algorithms including greedy algorithms, divide-and-conquer strategies, and dynamic programming.

This chapter is divided into five sections. It starts with the definition of algorithms and their complexity in Section 2.1. We introduce the asymptotic O-notation used in the analysis of the running time and space of an algorithm. Two tables are used to demonstrate that the asymptotic complexity of an algorithm will ultimately determine the size of problems that can be solved by the algorithm.

Then, we introduce greedy algorithms in Section 2.2. For some optimization problems, greedy algorithms are more efficient. A greedy algorithm pursues the best choice at the moment in the hope that it will lead to the best solution in the end. It works quite well for a wide range of problems. Huffman's algorithm is used as an example of a greedy algorithm.

Section 2.3 describes another common algorithmic technique, called divide-and-conquer. This strategy divides the problem into smaller parts, conquers each part individually, and then combines them to form a solution for the whole. We use the mergesort algorithm to illustrate the divide-and-conquer algorithm design paradigm.

Following its introduction by Needleman and Wunsch, dynamic programming has become a major algorithmic strategy for many optimization problems in sequence comparison. The development of a dynamic-programming algorithm has three basic components: the recurrence relation for defining the value of an optimal solution, the tabular computation for computing the value of an optimal solution, and the backtracking procedure for delivering an optimal solution. In Section 2.4, we introduce these basic ideas by developing dynamic-programming solutions for problems from different application areas, including the maximum-sum segment problem, the longest increasing subsequence problem, and the longest common subsequence problem.

Finally, we conclude the chapter with the bibliographic notes in Section 2.5.

2.1 Algorithms and Their Complexity

An *algorithm* is a step-by-step procedure for solving a problem by a computer. When an algorithm is executed by a computer, the central processing unit (CPU) performs the operations and the memory stores the program and data.

Let n be the size of the input, the output, or their sum. The time or space complexity of an algorithm is usually denoted as a function $f(n)$. Table 2.1 calculates the time needed if the function stands for the number of operations required by an algorithm, and we assume that the CPU performs one million operations per second.

Exponential algorithms grow pretty fast and become impractical even when n is small. For those quadratic and cubic functions, they grow faster than the linear functions. The constant and log factor matter, but are mostly acceptable in practice. As a rule of thumb, algorithms with a quadratic time complexity or higher are often impractical for large data sets.

Table 2.2 further shows the growth of the input size solvable by polynomial and exponential time algorithms with improved computers. Even with a million-times faster computer, the 10^n algorithm only adds 6 to the input size, which makes it hopeless for handling a moderate-size input.

These observations lead to the definition of the O-notation, which is very useful for the analysis of algorithms. We say $f(n) = O(g(n))$ if and only if there exist two positive constants c and n_0 such that $0 \leq f(n) \leq cg(n)$ for all $n \geq n_0$. In other words, for sufficiently large n, $f(n)$ can be bounded by $g(n)$ times a constant. In this kind of asymptotic analysis, the most crucial part is the order of the function, not the constant. For example, if $f(n) = 3n^2 + 5n$, we can say $f(n) = O(n^2)$ by letting $c = 4$ and $n_0 = 10$. By definition, it is also correct to say $n^2 = O(n^3)$, but we always prefer to choose a tighter order if possible. On the other hand, $10^n \neq O(n^x)$ for any integer x. That is, an exponential function cannot be bounded above by any polynomial function.

2.2 Greedy Algorithms

A greedy method works in stages. It always makes a locally optimal (*greedy*) choice at each stage. Once a choice has been made, it cannot be withdrawn, even if later we

Table 2.1 The time needed by the functions where we assume one million operations per second.

$f(n)$	$n = 10$	$n = 100$	$n = 100000$
$30n$	0.0003 second	0.003 second	3 seconds
$100n \log_{10} n$	0.001 second	0.02 second	50 seconds
$3n^2$	0.0003 second	0.03 second	8.33 hours
n^3	0.001 second	1 second	31.71 years
10^n	2.78 hours	3.17×10^{84} centuries	3.17×10^{99984} centuries

Table 2.2 The growth of the input size solvable in an hour as the computer runs faster.

$f(n)$	Present speed	1000-times faster	10^6-times faster
n	x_1	$1000x_1$	$10^6 x_1$
n^2	x_2	$31.62x_2$	$10^3 x_2$
n^3	x_3	$10x_3$	$10^2 x_3$
10^n	x_4	$x_4 + 3$	$x_4 + 6$

realize that it is a poor decision. In other words, this greedy choice may or may not lead to a globally optimal solution, depending on the characteristics of the problem.

It is a very straightforward algorithmic technique and has been used to solve a variety of problems. In some situations, it is used to solve the problem exactly. In others, it has been proved to be effective in approximation.

What kind of problems are suitable for a greedy solution? There are two ingredients for an optimization problem to be exactly solved by a greedy approach. One is that it has the so-called greedy-choice property, meaning that a locally optimal choice can reach a globally optimal solution. The other is that it satisfies the principle of optimality, *i.e.*, each solution substructure is optimal. We use Huffman coding, a frequency-dependent coding scheme, to illustrate the greedy approach.

2.2.1 Huffman Codes

Suppose we are given a very long DNA sequence where the occurrence probabilities of nucleotides A (adenine), C (cytosine), G (guanine), T (thymine) are $0.1, 0.1, 0.3$, and 0.5, respectively. In order to store it in a computer, we need to transform it into a binary sequence, using only 0's and 1's. A trivial solution is to encode A, C, G, and T by "00," "01," "10," and "11," respectively. This representation requires two bits per nucleotide. The question is "Can we store the sequence in a more compressed way?" Fortunately, by assigning longer codes for frequent nucleotides G and T, and shorter codes for rare nucleotides A and C, it can be shown that it requires less than two bits per nucleotide on average.

In 1952, Huffman [94] proposed a greedy algorithm for building up an optimal way of representing each letter as a binary string. It works in two phases. In phase one, we build a binary tree based on the occurrence probabilities of the letters. To do so, we first write down all the letters, together with their associated probabilities. They are initially the unmarked terminal nodes of the binary tree that we will build up as the algorithm proceeds. As long as there is more than one unmarked node left, we repeatedly find the two unmarked nodes with the smallest probabilities, mark them, create a new unmarked internal node with an edge to each of the nodes just marked, and set its probability as the sum of the probabilities of the two nodes.

The tree building process is depicted in Figure 2.1. Initially, there are four unmarked nodes with probabilities $0.1, 0.1, 0.3$, and 0.5. The two smallest ones are

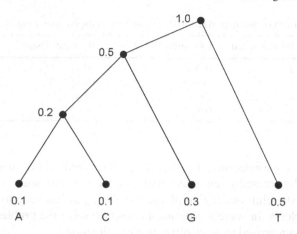

Fig. 2.1 Building a binary tree based on the occurrence probabilities of the letters.

with probabilities 0.1 and 0.1. Thus we mark these two nodes and create a new node with probability 0.2 and connect it to the two nodes just marked. Now we have three unmarked nodes with probabilities 0.2, 0.3, and 0.5. The two smallest ones are with probabilities 0.2 and 0.3. They are marked and a new node connecting them with probabilities 0.5 is created. The final iteration connects the only two unmarked nodes with probabilities 0.5 and 0.5. Since there is only one unmarked node left, *i.e.*, the root of the tree, we are done with the binary tree construction.

After the binary tree is built in phase one, the second phase is to assign the binary strings to the letters. Starting from the root, we recursively assign the value "zero" to the left edge and "one" to the right edge. Then for each leaf, *i.e.*, the letter, we concatenate the 0's and 1's from the root to it to form its binary string representation. For example, in Figure 2.2 the resulting codewords for A, C, G, and T are "000," "000," "01," and "1," respectively. By this coding scheme, a 20-nucleotide DNA sequence "GTTGTTATCGTTTATGTGGC" will be represented as a 34-bit binary sequence "0111011100010010111100010110101001." In general, since $3 \times 0.1 + 3 \times 0.1 + 2 \times 0.3 + 1 \times 0.5 = 1.7$, we conclude that, by Huffman coding techniques, each nucleotide requires 1.7 bits on average, which is superior to 2 bits by a trivial solution. Notice that in a Huffman code, no codeword is also a prefix of any other codeword. Therefore we can decode a binary sequence without any ambiguity. For example, if we are given "0111011100010010111100010110101001," we decode the binary sequence as "01" (G), "1" (T), "1" (T), "01" (G), and so forth.

The correctness of Huffman's algorithm lies in two properties: (1) greedy-choice property and (2) optimal-substructure property. It can be shown that there exists an optimal binary code in which the codewords for the two smallest-probability nodes have the same length and differ only in the last bit. That's the reason why we can contract them greedily without missing the path to the optimal solution. Besides, after contraction, the optimal-substructure property allows us to consider only those unmarked nodes.

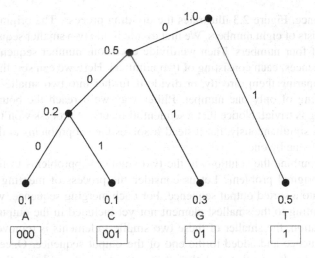

Fig. 2.2 Huffman code assignment.

Let n be the number of letters under consideration. For DNA, n is 4 and for English, n is 26. Since a heap can be used to maintain the minimum dynamically in $O(\log n)$ time for each insertion or deletion, the time complexity of Huffman's algorithm is $O(n \log n)$.

2.3 Divide-and-Conquer Strategies

The divide-and-conquer strategy *divides* the problem into a number of smaller subproblems. If the subproblem is small enough, it *conquers* it directly. Otherwise, it *conquers* the subproblem recursively. Once the solution to each subproblem has been done, it combines them together to form a solution to the original problem.

One of the well-known applications of the divide-and-conquer strategy is the design of sorting algorithms. We use mergesort to illustrate the divide-and-conquer algorithm design paradigm.

2.3.1 Mergesort

Given a sequence of n numbers $\langle a_1, a_2, \ldots, a_n \rangle$, the sorting problem is to sort these numbers into a nondecreasing sequence. For example, if the given sequence is $\langle 65, 16, 25, 85, 12, 8, 36, 77 \rangle$, then its sorted sequence is $\langle 8, 12, 16, 25, 36, 65, 77, 85 \rangle$.

To sort a given sequence, mergesort splits the sequence into half, sorts each of them recursively, then combines the resulting two sorted sequences into one sorted sequence. Figure 2.3 illustrates the dividing process. The original input sequence consists of eight numbers. We first divide it into two smaller sequences, each consisting of four numbers. Then we divide each four-number sequence into two smaller sequences, each consisting of two numbers. Here we can sort the two numbers by comparing them directly, or divide it further into two smaller sequences, each consisting of only one number. Either way we'll reach the boundary cases where sorting is trivial. Notice that a sequential recursive process won't expand the subproblems simultaneously, but instead it solves the subproblems at the same recursion depth one by one.

How to combine the solutions to the two smaller subproblems to form a solution to the original problem? Let us consider the process of merging two sorted sequences into a sorted output sequence. For each merging sequence, we maintain a cursor pointing to the smallest element not yet included in the output sequence. At each iteration, the smaller of these two smallest elements is removed from the merging sequence and added to the end of the output sequence. Once one merging sequence has been exhausted, the other sequence is appended to the end of the output sequence. Figure 2.4 depicts the merging process. The merging sequences are $\langle 16, 25, 65, 85 \rangle$ and $\langle 8, 12, 36, 77 \rangle$. The smallest elements of the two merging sequences are 16 and 8. Since 8 is a smaller one, we remove it from the merging sequence and add it to the output sequence. Now the smallest elements of the two merging sequences are 16 and 12. We remove 12 from the merging sequence and append it to the output sequence. Then 16 and 36 are the smallest elements of the two merging sequences, thus 16 is appended to the output list. Finally, the resulting output sequence is $\langle 8, 12, 16, 25, 36, 65, 77, 85 \rangle$. Let N and M be the lengths of the

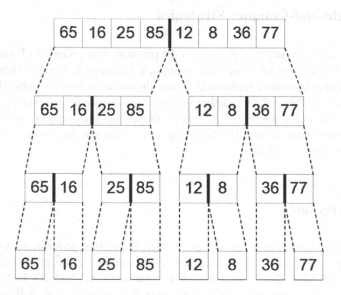

Fig. 2.3 The top-down dividing process of mergesort.

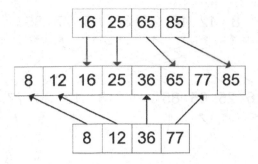

Fig. 2.4 The merging process of mergesort.

two merging sequences. Since the merging process scans the two merging sequences linearly, its running time is therefore $O(N + M)$ in total.

After the top-down dividing process, mergesort accumulates the solutions in a bottom-up fashion by combining two smaller sorted sequences into a larger sorted sequence as illustrated in Figure 2.5. In this example, the recursion depth is $\lceil \log_2 8 \rceil = 3$. At recursion depth 3, every single element is itself a sorted sequence. They are merged to form sorted sequences at recursion depth 2: $\langle 16, 65 \rangle$, $\langle 25, 85 \rangle$, $\langle 8, 12 \rangle$, and $\langle 36, 77 \rangle$. At recursion depth 1, they are further merged into two sorted sequences: $\langle 16, 25, 65, 85 \rangle$ and $\langle 8, 12, 36, 77 \rangle$. Finally, we merge these two sequences into one sorted sequence: $\langle 8, 12, 16, 25, 36, 65, 77, 85 \rangle$.

It can be easily shown that the recursion depth of mergesort is $\lceil \log_2 n \rceil$ for sorting n numbers, and the total time spent for each recursion depth is $O(n)$. Thus, we conclude that mergesort sorts n numbers in $O(n \log n)$ time.

2.4 Dynamic Programming

Dynamic programming is a class of solution methods for solving sequential decision problems with a compositional cost structure. It is one of the major paradigms of algorithm design in computer science. Like the usage in *linear programming*, the word "*programming*" refers to *finding an optimal plan* of action, rather than *writing programs*. The word "*dynamic*" in this context conveys the idea that choices may depend on the current state, rather than being decided ahead of time.

Typically, dynamic programming is applied to optimization problems. In such problems, there exist many possible solutions. Each solution has a value, and we wish to find a solution with the optimum value. There are two ingredients for an optimization problem to be suitable for a dynamic-programming approach. One is that it satisfies the principle of optimality, *i.e.*, each solution substructure is optimal. Greedy algorithms require this very same ingredient, too. The other ingredient is that it has overlapping subproblems, which has the implication that it can be solved

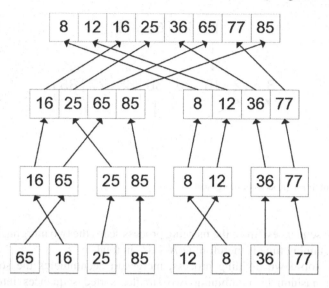

Fig. 2.5 Accumulating the solutions in a bottom-up manner.

more efficiently if the solutions to the subproblems are recorded. If the subproblems are not overlapping, a divide-and-conquer approach is the choice.

The development of a dynamic-programming algorithm has three basic components: the recurrence relation for defining the value of an optimal solution, the tabular computation for computing the value of an optimal solution, and the back-tracking procedure for delivering an optimal solution. Here we introduce these basic ideas by developing dynamic-programming solutions for problems from different application areas.

First of all, the Fibonacci numbers are used to demonstrate how a tabular computation can avoid recomputation. Then we use three classic problems, namely, the maximum-sum segment problem, the longest increasing subsequence problem, and the longest common subsequence problem, to explain how dynamic-programming approaches can be used to solve the sequence-related problems.

2.4.1 Fibonacci Numbers

The Fibonacci numbers were first created by Leonardo Fibonacci in 1202. It is a simple series, but its applications are nearly everywhere in nature. It has fascinated mathematicians for more than 800 years. The *Fibonacci numbers* are defined by the following recurrence:

$$\begin{cases} F_0 = 0, \\ F_1 = 1, \\ F_i = F_{i-1} + F_{i-2} \text{ for } i \geq 2. \end{cases}$$

By definition, the sequence goes like this: 0, 1, 1, 2, 3, 5, 8, 13, 21, 34, 55, 89, 144, 233, 377, 610, 987, 1597, 2584, 4181, 6765, 10946, 17711, 28657, 46368, 75025, 121393, and so forth. Given a positive integer n, how would you compute F_n? You might say that it can be easily solved by a straightforward divide-and-conquer method based on the recurrence. That's right. But is it efficient? Take the computation of F_{10} for example (see Figure 2.6). By definition, F_{10} is derived by adding up F_9 and F_8. What about the values of F_9 and F_8? Again, F_9 is derived by adding up F_8 and F_7; F_8 is derived by adding up F_7 and F_6. Working toward this direction, we'll finally reach the values of F_1 and F_0, i.e., the end of the recursive calls. By adding them up backwards, we have the value of F_{10}. It can be shown that the number of recursive calls we have to make for computing F_n is exponential in n.

Those who are ignorant of history are doomed to repeat it. A major drawback of this divide-and-conquer approach is to solve many of the subproblems repeatedly. A tabular method solves every subproblem just once and then saves its answer in a table, thereby avoiding the work of recomputing the answer every time the subproblem is encountered. Figure 2.7 explains that F_n can be computed in $O(n)$ steps by a tabular computation. It should be noted that F_n can be computed in just $O(\log n)$ steps by applying matrix computation.

2.4.2 The Maximum-Sum Segment Problem

Given a sequence of numbers $A = \langle a_1, a_2, \ldots, a_n \rangle$, the maximum-sum segment problem is to find, in A, a consecutive subsequence, i.e., a substring or segment, with the maximum sum. For each position i, we can compute the maximum-sum segment ending at that position in $O(i)$ time. Therefore, a naive algorithm runs in $\sum_{i=1}^{n} O(i) = O(n^2)$ time.

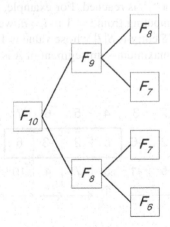

Fig. 2.6 Computing F_{10} by divide-and-conquer.

F_0	F_1	F_2	F_3	F_4	F_5	F_6	F_7	F_8	F_9	F_{10}
0	1	1	2	3	5	8	13	21	34	55

Fig. 2.7 Computing F_{10} by a tabular computation.

Now let us describe a more efficient dynamic-programming algorithm for this problem. Define $S[i]$ to be the maximum sum of segments ending at position i of A. The value $S[i]$ can be computed by the following recurrence:

$$S[i] = \begin{cases} a_i + \max\{S[i-1], 0\} & \text{if } i > 1, \\ a_1 & \text{if } i = 1. \end{cases}$$

If $S[i-1] < 0$, concatenating a_i with its previous elements will give a smaller sum than a_i itself. In this case, the maximum-sum segment ending at position i is a_i itself.

By a tabular computation, each $S[i]$ can be computed in constant time for i from 1 to n, therefore all S values can be computed in $O(n)$ time. During the computation, we record the largest S entry computed so far in order to report where the maximum-sum segment ends. We also record the traceback information for each position i so that we can trace back from the end position of the maximum-sum segment to its start position. If $S[i-1] > 0$, we need to concatenate with previous elements for a larger sum, therefore the traceback symbol for position i is "←." Otherwise, "↑" is recorded. Once we have computed all S values, the traceback information is used to construct the maximum-sum segment by starting from the largest S entry and following the arrows until a "↑" is reached. For example, in Figure 2.8, $A = \langle 3, 2, -6, 5, 2, -3, 6, -4, 2 \rangle$. By computing from $i = 1$ to $i = n$, we have $S = \langle 3, 5, -1, 5, 7, 4, 10, 6, 8 \rangle$. The maximum S entry is $S[7]$ whose value is 10. By backtracking from $S[7]$, we conclude that the maximum-sum segment of A is $\langle 5, 2, -3, 6 \rangle$, whose sum is 10.

i	1	2	3	4	5	6	7	8	9
A	3	2	-6	5	2	-3	6	-4	2
S	3	5	-1	5	7	4	10	6	8
	↑	←	←	↑	←	←	←	←	←

Fig. 2.8 Finding a maximum-sum segment.

Let *prefix sum* $P[i] = \sum_{j=1}^{i} a_j$ be the sum of the first i elements. It can be easily seen that $\sum_{k=i}^{j} a_k = P[j] - P[i-1]$. Therefore, if we wish to compute for a given position the maximum-sum segment ending at it, we could just look for a minimum prefix sum ahead of this position. This yields another linear-time algorithm for the maximum-sum segment problem.

2.4.3 Longest Increasing Subsequences

Given a sequence of numbers $A = \langle a_1, a_2, \ldots, a_n \rangle$, the longest increasing subsequence problem is to find an increasing subsequence in A whose length is maximum. Without loss of generality, we assume that these numbers are distinct. Formally speaking, given a sequence of distinct real numbers $A = \langle a_1, a_2, \ldots, a_n \rangle$, sequence $B = \langle b_1, b_2, \ldots, b_k \rangle$ is said to be a subsequence of A if there exists a strictly increasing sequence $\langle i_1, i_2, \ldots, i_k \rangle$ of indices of A such that for all $j = 1, 2, \ldots, k$, we have $a_{i_j} = b_j$. In other words, B is obtained by deleting zero or more elements from A. We say that the subsequence B is increasing if $b_1 < b_2 < \ldots < b_k$. The longest increasing subsequence problem is to find a maximum-length increasing subsequence of A.

For example, suppose $A = \langle 4, 8, 2, 7, 3, 6, 9, 1, 10, 5 \rangle$, both $\langle 2, 3, 6 \rangle$ and $\langle 2, 7, 9, 10 \rangle$ are increasing subsequences of A, whereas $\langle 8, 7, 9 \rangle$ (not increasing) and $\langle 2, 3, 5, 7 \rangle$ (not a subsequence) are not.

Note that we may have more than one longest increasing subsequence, so we use "*a* longest increasing subsequence" instead of "*the* longest increasing subsequence." Let $L[i]$ be the length of a longest increasing subsequence ending at position i. They can be computed by the following recurrence:

$$L[i] = \begin{cases} 1 + \max_{j=0,\ldots,i-1}\{L[j] \mid a_j < a_i\} & \text{if } i > 0, \\ 0 & \text{if } i = 0. \end{cases}$$

Here we assume that a_0 is a dummy element and smaller than any element in A, and $L[0]$ is equal to 0. By tabular computation for every i from 1 to n, each $L[i]$ can be computed in $O(i)$ steps. Therefore, they require in total $\sum_{i=1}^{n} O(i) = O(n^2)$ steps. For each position i, we use an array P to record the index of the best previous element for the current element to concatenate with. By tracing back from the element with the largest L value, we derive a longest increasing subsequence.

Figure 2.9 illustrates the process of finding a longest increasing subsequence of $A = \langle 4, 8, 2, 7, 3, 6, 9, 1, 10, 5 \rangle$. Take $i = 4$ for instance, where $a_4 = 7$. Its previous smaller elements are a_1 and a_3, both with L value equaling 1. Therefore, we have $L[4] = L[1] + 1 = 2$, meaning that the length of a longest increasing subsequence ending at position 4 is of length 2. Indeed, both $\langle a_1, a_4 \rangle$ and $\langle a_3, a_4 \rangle$ are an increasing subsequence ending at position 4. In order to trace back the solution, we use array P to record which entry contributes the maximum to the current L value. Thus, $P[4]$ can be 1 (standing for a_1) or 3 (standing for a_3). Once we have computed all L and

P values, the maximum L value is the length of a longest increasing subsequence of A. In this example, $L[9] = 5$ is the maximum. Tracing back from $P[9]$, we have found a longest increasing subsequence $\langle a_3, a_5, a_6, a_7, a_9 \rangle$, i.e., $\langle 2, 3, 6, 9, 10 \rangle$.

In the following, we briefly describe a more efficient dynamic-programming algorithm for delivering a longest increasing subsequence. A crucial observation is that it suffices to store only those smallest ending elements for all possible lengths of the increasing subsequences. For example, in Figure 2.9, there are three entries whose L value is 2, namely $a_2 = 8$, $a_4 = 7$, and $a_5 = 3$, where a_5 is the smallest. Any element after position 5 that is larger than a_2 or a_4 is also larger than a_5. Therefore, a_5 can replace the roles of a_2 and a_4 after position 5.

Let $SmallestEnd[k]$ denote the smallest ending element of all possible increasing subsequences of length k ending before the current position i. The algorithm proceeds for i from 1 to n. How do we update $SmallestEnd[k]$ when we consider a_i? By definition, it is easy to see that the elements in $SmallestEnd$ are in increasing order. In fact, a_i will affect only one entry in $SmallestEnd$. If a_i is larger than all the elements in $SmallestEnd$, then we can concatenate a_i to the longest increasing subsequence computed so far. That is, one more entry is added to the end of $SmallestEnd$. A backtracking pointer is recorded by pointing to the previous last element of $SmallestEnd$. Otherwise, let $SmallestEnd[k']$ be the smallest element that is larger than a_i. We replace $SmallestEnd[k']$ by a_i because now we have a smaller ending element of an increasing subsequence of length k'.

Since $SmallestEnd$ is a sorted array, the above process can be done by a binary search. A binary search algorithm compares the query element with the middle element of the sorted array, if the query element is larger, then it searches the larger half recursively. Otherwise, it searches the smaller half recursively. Either way the size of the search space is shrunk by a factor of two. At position i, the size of $SmallestEnd$ is at most i. Therefore, for each position i, it takes $O(\log i)$ time to determine the appropriate entry to be updated by a_i. Therefore, in total we have an $O(n \log n)$-time algorithm for the longest increasing subsequence problem.

Figure 2.10 illustrates the process of finding a longest increasing subsequence of $A = \langle 4, 8, 2, 7, 3, 6, 9, 1, 10, 5 \rangle$. When $i = 1$, there is only one increasing subsequence, i.e., $\langle 4 \rangle$. We have $SmallestEnd[1] = 4$. Since $a_2 = 8$ is larger than $SmallestEnd[1]$, we create a new entry $SmallestEnd[2] = 8$ and set the backtracking

i	1	2	3	4	5	6	7	8	9	10
A	4	8	2	7	3	6	9	1	10	5
L	1	2	1	2	2	3	4	1	5	3
P	0	1	0	1	3	5	6	0	7	5

Fig. 2.9 An $O(n^2)$-time algorithm for finding a longest increasing subsequence.

pointer $P[2] = 1$, meaning that a_2 can be concatenated with a_1 to form an increasing subsequence $\langle 4, 8 \rangle$. When $a_3 = 2$ is encountered, its nearest larger element in *SmallestEnd* is $SmallestEnd[1] = 4$. We know that we now have an increasing subsequence $\langle 2 \rangle$ of length 1. So $SmallestEnd[1]$ is changed from 4 to $a_3 = 2$ and $P[3] = 0$. When $i = 4$, we have $SmallestEnd[1] = 2 < a_4 = 7 < SmallestEnd[2] = 8$. By concatenating a_4 with $Smallest[1]$, we have a new increasing subsequence $\langle 2, 7 \rangle$ of length 2 whose ending element is smaller than 8. Thus, $SmallestEnd[2]$ is changed from 8 to $a_4 = 7$ and $P[4] = 3$. Continue this way until we reach a_{10}. When a_{10} is encountered, we have $SmallestEnd[2] = 3 < a_{10} = 5 < SmallestEnd[3] = 6$. We set $SmallestEnd[3] = a_{10} = 5$ and $P[10] = 5$. Now the largest element in *SmallestEnd* is $SmallestEnd[5] = a_9 = 10$. We can trace back from a_9 by the backtracking pointers P and deliver a longest increasing subsequence $\langle a_3, a_5, a_6, a_7, a_9 \rangle$, i.e., $\langle 2, 3, 6, 9, 10 \rangle$.

2.4.4 Longest Common Subsequences

A subsequence of a sequence S is obtained by deleting zero or more elements from S. For example, $\langle P, R, E, D \rangle$, $\langle S, D, N \rangle$, and $\langle P, R, E, D, E, N, T \rangle$ are all subsequences of $\langle P, R, E, S, I, D, E, N, T \rangle$, whereas $\langle S, N, D \rangle$ and $\langle P, E, F \rangle$ are not.

Recall that, given two sequences, the longest common subsequence (LCS) problem is to find a subsequence that is common to both sequences and its length is maximized. For example, given two sequences

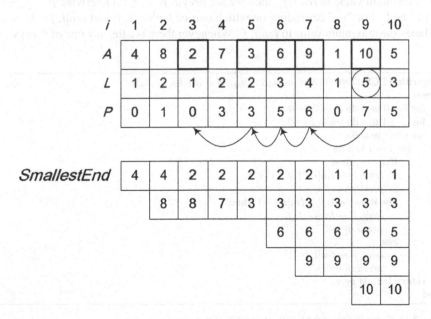

Fig. 2.10 An $O(n \log n)$-time algorithm for finding a longest increasing subsequence.

$$\langle P,R,E,S,I,D,E,N,T \rangle$$

and

$$\langle P,R,O,V,I,D,E,N,C,E \rangle,$$

$\langle P,R,D,N \rangle$ is a common subsequence of them, whereas $\langle P,R,V \rangle$ is not. Their LCS is $\langle P,R,I,D,E,N \rangle$.

Now let us formulate the recurrence for computing the length of an LCS of two sequences. We are given two sequences $A = \langle a_1,a_2,\ldots,a_m \rangle$, and $B = \langle b_1,b_2,\ldots,b_n \rangle$. Let $len[i,j]$ denote the length of an LCS between $\langle a_1,a_2,\ldots,a_i \rangle$ (a prefix of A) and $\langle b_1,b_2,\ldots,b_j \rangle$ (a prefix of B). They can be computed by the following recurrence:

$$len[i,j] = \begin{cases} 0 & \text{if } i = 0 \text{ or } j = 0, \\ len[i-1,j-1]+1 & \text{if } i,j > 0 \text{ and } a_i = b_j, \\ \max\{len[i,j-1], len[i-1,j]\} & \text{otherwise.} \end{cases}$$

In other words, if one of the sequences is empty, the length of their LCS is just zero. If a_i and b_j are the same, an LCS between $\langle a_1,a_2,\ldots,a_i \rangle$, and $\langle b_1,b_2,\ldots,b_j \rangle$ is the concatenation of an LCS of $\langle a_1,a_2,\ldots,a_{i-1} \rangle$ and $\langle b_1,b_2,\ldots,b_{j-1} \rangle$ and a_i. Therefore, $len[i,j] = len[i-1,j-1]+1$ in this case. If a_i and b_j are different, their LCS is equal to either an LCS of $\langle a_1,a_2,\ldots,a_i \rangle$, and $\langle b_1,b_2,\ldots,b_{j-1} \rangle$, or that of $\langle a_1,a_2,\ldots,a_{i-1} \rangle$, and $\langle b_1,b_2,\ldots,b_j \rangle$. Its length is thus the maximum of $len[i,j-1]$ and $len[i-1,j]$.

Figure 2.11 gives the pseudo-code for computing $len[i,j]$. For each entry (i,j), we retain the backtracking information in $prev[i,j]$. If $len[i-1,j-1]$ contributes the maximum value to $len[i,j]$, then we set $prev[i,j] = $"$\searmark$." Otherwise $prev[i,j]$ is set to be "$\uparrow$" or "$\leftarrow$" depending on which one of $len[i-1,j]$ and $len[i,j-1]$ contributes the maximum value to $len[i,j]$. Whenever there is a tie, any one of them will

---

Algorithm LCS_LENGTH($A = \langle a_1,a_2,\ldots,a_m \rangle$, $B = \langle b_1,b_2,\ldots,b_n \rangle$)
begin
    for $i \leftarrow 0$ **to** m **do** $len[i,0] \leftarrow 0$
    for $j \leftarrow 1$ **to** n **do** $len[0,j] \leftarrow 0$
    for $i \leftarrow 1$ **to** m **do**
        for $j \leftarrow 1$ **to** n **do**
            if $a_i = b_j$ **then**
                $len[i,j] \leftarrow len[i-1,j-1]+1$
                $prev[i,j] \leftarrow$ "$\searrow$"
            else if $len[i-1,j] \geq len[i,j-1]$ **then**
                $len[i,j] \leftarrow len[i-1,j]$
                $prev[i,j] \leftarrow$ "$\uparrow$"
            else
                $len[i,j] \leftarrow len[i,j-1]$
                $prev[i,j] \leftarrow$ "$\leftarrow$"
    return len and $prev$
end

Fig. 2.11 Computation of the length of an LCS of two sequences.

		A	L	I	G	N	M	E	N	T
	0	0	0	0	0	0	0	0	0	0
A	0	↖1	←1	←1	←1	←1	←1	←1	←1	←1
L	0	↑1	↖2	←2	←2	←2	←2	←2	←2	←2
G	0	↑1	↑2	↑2	↖3	←3	←3	←3	←3	←3
O	0	↑1	↑2	↑2	↑3	↑3	↑3	↑3	↑3	↑3
R	0	↑1	↑2	↑2	↑3	↑3	↑3	↑3	↑3	↑3
I	0	↑1	↑2	↖3	↑3	↑3	↑3	↑3	↑3	↑3
T	0	↑1	↑2	↑3	↑3	↑3	↑3	↑3	↑3	↖4
H	0	↑1	↑2	↑3	↑3	↑3	↑3	↑3	↑3	↑4
M	0	↑1	↑2	↑3	↑3	↑3	↖4	←4	←4	↑4

Fig. 2.12 Tabular computation of the length of an LCS of $\langle A,L,G,O,R,I,T,H,M \rangle$ and $\langle A,L,I,G,N,M,E,N,T \rangle$.

work. These arrows will guide the backtracking process upon reaching the terminal entry (m,n). Since the time spent for each entry is $O(1)$, the total running time of algorithm LCS_LENGTH is $O(mn)$.

Figure 2.12 illustrates the tabular computation. The length of an LCS of

$$\langle A,L,G,O,R,I,T,H,M \rangle$$

and

$$\langle A,L,I,G,N,M,E,N,T \rangle$$

is 4.

Besides computing the length of an LCS of the whole sequences, Figure 2.12 in fact computes the length of an LCS between each pair of prefixes of the two sequences. For example, by this table, we can also tell the length of an LCS between $\langle A,L,G,O,R \rangle$ and $\langle A,L,I,G \rangle$ is 3.

Algorithm LCS_OUTPUT($A = \langle a_1, a_2, \ldots, a_m \rangle$, *prev*, i, j)
begin
    if $i = 0$ or $j = 0$ **then return**
    if $prev[i,j] =$ "↖" **then**
        LCS_OUTPUT(A, *prev*, $i-1$, $j-1$)
        print a_i
    else if $prev[i,j] =$ "↑" **then** LCS_OUTPUT(A, *prev*, $i-1$, j)
    else LCS_OUTPUT(A, *prev*, i, $j-1$)
end

Fig. 2.13 Delivering an LCS.

Once algorithm LCS_LENGTH reaches (m,n), the backtracking information retained in array *prev* allows us to find out which common subsequence contributes $len[m,n]$, the maximum length of an LCS of sequences A and B. Figure 2.13 lists the pseudo-code for delivering an LCS. We trace back the dynamic-programming matrix from the entry (m,n) recursively following the direction of the arrow. Whenever a diagonal arrow "↖" is encountered, we append the current matched letter to the end of the LCS under construction. Algorithm LCS_OUTPUT takes $O(m+n)$ time in total since each recursive call reduces the indices i and/or j by one.

Figure 2.14 backtracks the dynamic-programming matrix computed in Figure 2.12. It outputs $\langle A,L,G,T \rangle$ (the shaded entries) as an LCS of

$$\langle A,L,G,O,R,I,T,H,M \rangle$$

and

$$\langle A,L,I,G,N,M,E,N,T \rangle.$$

		(A)	(L)	I	(G)	N	M	E	N	(T)
	0	0	0	0	0	0	0	0	0	0
(A)	0	↖1	←1	←1	←1	←1	←1	←1	←1	←1
(L)	0	↑1	↖2	←2	←2	←2	←2	←2	←2	←2
(G)	0	↑1	↑2	↑2	↖3	←3	←3	←3	←3	←3
O	0	↑1	↑2	↑2	↑3	↑3	↑3	↑3	↑3	↑3
R	0	↑1	↑2	↑2	↑3	↑3	↑3	↑3	↑3	↑3
I	0	↑1	↑2	↖3	↑3	↑3	↑3	↑3	↑3	↑3
(T)	0	↑1	↑2	↑3	↑3	↑3	↑3	↑3	↑3	↖4
H	0	↑1	↑2	↑3	↑3	↑3	↑3	↑3	↑3	↑4
M	0	↑1	↑2	↑3	↑3	↑3	↖4	←4	←4	↑4

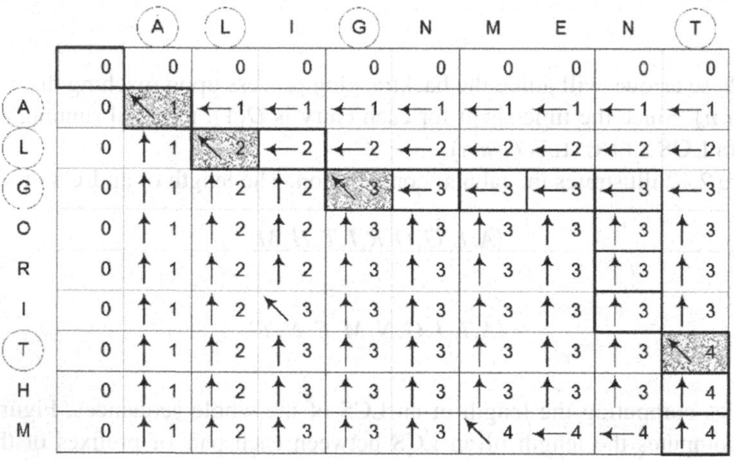

Fig. 2.14 Backtracking process for finding an LCS of $\langle A,L,G,O,R,I,T,H,M \rangle$ and $\langle A,L,I,G,N,M,E,N,T \rangle$.

2.5 Bibliographic Notes and Further Reading

This chapter presents three basic algorithmic techniques that are often used in designing efficient methods for various problems in sequence comparison. Readers can refer to algorithm textbooks for more instructive tutorials. The algorithm book (or "The White Book") by Cormen et al. [52] is a comprehensive reference of data structures and algorithms with a solid mathematical and theoretical foundation. Manber's book [133] provides a creative approach for the design and analysis of algorithms. The book by Baase and Gelder [17] is a good algorithm textbook for beginners.

2.1

As noted by Donald E. Knuth [113], the invention of the O-notation originated from a number-theory book by P. Bachmann in 1892.

2.2

David A. Huffman [94] invented Huffman coding while he was a Ph.D. student at MIT. It was actually a term paper for the problem of finding the most efficient binary coding scheme assigned by Robert M. Fano.

2.3

There are numerous sorting algorithms such as insertion sort, bubblesort, quicksort, mergesort, to name a few. As noted by Donald E. Knuth [114], the first program ever written for a stored program computer was the mergesort program written by John von Neumann in 1945.

2.4

The name "dynamic programming" was given by Richard Bellman in 1957 [25]. The maximum-sum segment problem was first surveyed by Bentley and is linear-time solvable using Kadane's algorithm [27].

Chapter 3
Pairwise Sequence Alignment

Pairwise alignment is often used to reveal similarities between sequences, determine the residue-residue correspondences, locate patterns of conservation, study gene regulation, and infer evolutionary relationships.

This chapter is divided into eight sections. It starts with a brief introduction in Section 3.1, followed by the dot matrix representation of pairwise sequence comparison in Section 3.2.

Using alignment graph, we derive a dynamic-programming method for aligning globally two sequences in Section 3.3. An example is used to illustrate the tabular computation for computing the optimal alignment score as well as the backtracking procedure for delivering an optimal global alignment.

Section 3.4 describes a method for delivering an optimal local alignment, which involves only a segment of each sequence. The recurrence for an optimal local alignment is quite similar to that for global alignment. To reflect the flexibility that an alignment can start at any position of the two sequences, an additional entry "zero" is added.

In Section 3.5, we address some flexible strategies for coping with various scoring schemes. Affine gap penalties are considered more appropriate for aligning DNA and protein sequences. To favor longer gaps, constant gap penalties or restricted affine gap penalties could be the choice.

Straightforward implementations of the dynamic-programming algorithms consume quadratic space for alignment. For certain applications, such as careful analysis of a few long DNA sequences, the space restriction is more important than the time constraint. Section 3.6 introduces Hirschberg's linear-space approach.

Section 3.7 discusses several advanced topics such as constrained sequence alignment, similar sequence alignment, suboptimal alignment, and robustness measurement.

Finally, we conclude the chapter with the bibliographic notes in Section 3.8.

3.1 Introduction

In nature, even a single amino acid sequence contains all the information necessary
to determine the fold of the protein. However, the folding process is still mysterious
to us, and some valuable information can be revealed by sequence comparison. Take
a look at the following sequence:

THETR UTHIS MOREI MPORT ANTTH ANTHE FACTS

What did you see in the above sequence? By comparing it with the words in the
dictionary, we find the tokens "FACTS," "IMPORTANT," "IS," "MORE," "THAN,"
"THE," and "TRUTH." Then we figure out the above is the sentence "The truth is
more important than the facts."

Even though we have not yet decoded the DNA and protein languages, the emerging flood of sequence data has provided us with a golden opportunity of investigating
the evolution and function of biomolecular sequences. We are in a stage of compil-
ing dictionaries for DNA, proteins, and so forth. Sequence comparison plays a major
role in this line of research and thus becomes the most basic tool of bioinformatics.

Sequence comparison has wide applications to molecular biology, computer science, speech processing, and so on. In molecular biology, it is often used to reveal
similarities among sequences, determine the residue-residue correspondences, locate patterns of conservation, study gene regulation, and infer evolutionary relationships. It helps us to fish for related sequences in databanks, such as the GenBank
database. It can also be used for the annotation of genomes.

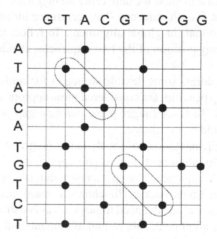

Fig. 3.1 A dot matrix of the two sequences ATACATGTCT and GTACGTCGG.

3.2 Dot Matrix

A dot matrix is a two-dimensional array of dots used to highlight the exact matches between two sequences. Given are two sequences $A = \langle a_1, a_2, \ldots, a_m \rangle$ (or $A = a_1 a_2 \ldots a_m$ in short), and $B = \langle b_1, b_2, \ldots, b_n \rangle$. A dot is plotted on the (i, j) entry of the matrix if $a_i = b_j$. Users can easily identify similar regions between the two sequences by locating those contiguous dots along the same diagonal. Figure 3.1 gives a dot matrix of the two sequences ATACATGTCT and GTACGTCGG. Dashed lines circle those regions with at lease three contiguous matches on the same diagonal.

A dot matrix allows the users to quickly visualize the similar regions of two sequences. However, as the sequences get longer, it becomes more involved to determine their most similar regions, which can no longer be answered by merely looking at a dot matrix. It would be more desirable to automatically identify those similar regions and rank them by their "similarity scores." This leads to the development of sequence alignment.

3.3 Global Alignment

Following its introduction by Needleman and Wunsch in 1970, dynamic programming has become a major algorithmic strategy for many optimization problems in sequence comparison. This strategy is guaranteed to produce an alignment of two given sequences having the highest score for a number of useful alignment-scoring schemes.

Given two sequences $A = a_1 a_2 \ldots a_m$, and $B = b_1 b_2 \ldots b_n$, an *alignment* of A and B is obtained by introducing dashes into the two sequences such that the lengths of the two resulting sequences are identical and no column contains two dashes. Let Σ denote the alphabet over which A and B are defined. To simplify the presentation, we employ a very simple scoring scheme as follows. A score $\sigma(a, b)$ is defined for each $(a, b) \in \Sigma \times \Sigma$. Each indel, *i.e.*, a column with a space, is penalized by a constant β. The score of an alignment is the sum of σ scores of all columns with no dashes minus the penalties of the gaps. An *optimal global alignment* is an alignment that

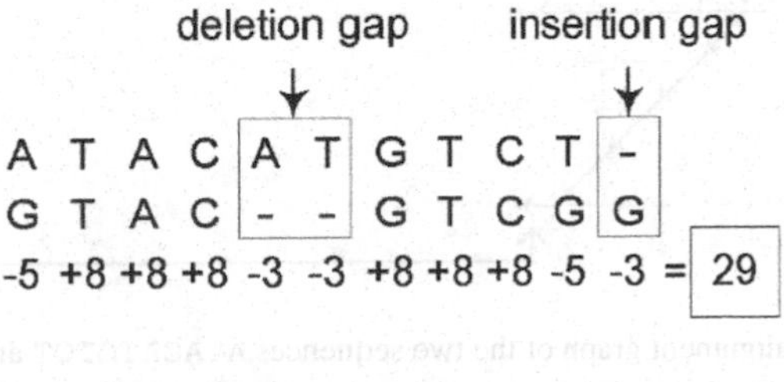

Fig. 3.2 An alignment of the two sequences ATACATGTCT and GTACGTCGG.

maximizes the score. By global alignment, we mean that both sequences are aligned globally, i.e., from their first symbols to their last.

Figure 3.2 gives an alignment of sequences ATACATGTCT and GTACGTCGG and its score. In this and the next sections, we assume the following simple scoring scheme. A match is given a bonus score 8, a mismatch is penalized by assigning score -5, and the gap penalty for each indel is -3. In other words, $\sigma(a,b) = 8$ if a and b are the same, $\sigma(a,b) = -5$ if a and b are different, and $\beta = -3$.

It is quite helpful to recast the problem of aligning two sequences as an equivalent problem of finding a maximum-scoring path in the alignment graph defined in Section 1.2.2, as has been observed by a number of researchers. Recall that the alignment graph of A and B is a directed acyclic graph whose vertices are the pairs (i,j) where $i \in \{0,1,2,\ldots,m\}$ and $j \in \{0,1,2,\ldots,n\}$. These vertices are arrayed in $m+1$ rows and $n+1$ columns. The edge set consists of three types of edges. The substitution aligned pairs, insertion aligned pairs, and deletion aligned pairs correspond to the diagonal edges, horizontal edges, and vertical edges, respectively. Specifically, a vertical edge from $(i-1,j)$ to (i,j), which corresponds to a deletion of a_i, is drawn for $i \in \{1,2,\ldots,m\}$ and $j \in \{0,1,2,\ldots,n\}$. A horizontal edge from $(i,j-1)$ to (i,j), which corresponds to an insertion of b_j, is drawn for $i \in \{0,1,2,\ldots,m\}$ and $j \in \{1,2,\ldots,n\}$. A diagonal edge from $(i-1,j-1)$ to (i,j), which corresponds to a substitution of a_i with b_j, is drawn for $i \in \{1,2,\ldots,m\}$ and $j \in \{1,2,\ldots,n\}$.

It has been shown that an alignment corresponds to a path from the leftmost cell of the first row to the rightmost cell of the last row in the alignment graph. Figure 3.3 gives another example of this correspondence.

Let $S[i,j]$ denote the score of an optimal alignment between $a_1 a_2 \ldots a_i$ and $b_1 b_2 \ldots b_j$. By definition, we have $S[0,0] = 0$, $S[i,0] = -\beta \times i$, and $S[0,j] = -\beta \times j$. With these initializations, $S[i,j]$ for $i \in \{1,2,\ldots,m\}$ and $j \in \{1,2,\ldots,n\}$ can be computed by the following recurrence.

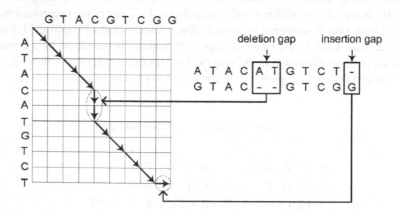

Fig. 3.3 A path in an alignment graph of the two sequences ATACATGTCT and GTACGTCGG.

$$S[i,j] = \max \begin{cases} S[i-1,j] - \beta, \\ S[i,j-1] - \beta, \\ S[i-1,j-1] + \sigma(a_i, b_j). \end{cases}$$

Figure 3.4 explains the recurrence by showing that there are three possible ways entering into the grid point (i,j), and we take the maximum of their path weights. The weight of the maximum-scoring path entering (i,j) from $(i-1,j)$ *vertically* is the weight of the maximum-scoring path entering $(i-1,j)$ plus the weight of edge $(i-1,j) \to (i,j)$. That is, the weight of the maximum-scoring path entering (i,j) with a deletion gap symbol at the end is $S[i-1,j] - \beta$. Similarly, the weight of the maximum-scoring path entering (i,j) from $(i,j-1)$ *horizontally* is $S[i,j-1] - \beta$ and the weight of the maximum-scoring path entering (i,j) from $(i-1,j-1)$ *diagonally* is $S[i-1,j-1] + \sigma(a_i, b_j)$. To compute $S[i,j]$, we simply take the maximum value of these three choices. The value $S[m,n]$ is the score of an optimal global alignment between sequences A and B.

Figure 3.5 gives the pseudo-code for computing the score of an optimal global alignment. Whenever there is a tie, any one of them will work. Since there are $O(mn)$ entries and the time spent for each entry is $O(1)$, the total running time of algorithm GLOBAL_ALIGNMENT_SCORE is $O(mn)$.

Now let us use an example to illustrate the tabular computation. Figure 3.6 computes the score of an optimal alignment of the two sequences ATACATGTCT and GTACGTCGG, where a match is given a bonus score 8, a mismatch is penalized by a score -5, and the gap penalty for each gap symbol is -3. The first row and column of the table are initialized with proper penalties. Other entries are computed in order. Take the entry $(5,3)$ for example. Upon computing the value of this entry, the following values are ready: $S[4,2] = -3$, $S[4,3] = 8$, and $S[5,2] = -6$. Since the edge weight of $(4,2) \to (5,3)$ is 8 (a match symbol "A"), the maximum score from $(4,2)$ to $(5,3)$ is $-3+8=5$. The maximum score from $(4,3)$ is $8-3=5$, and the

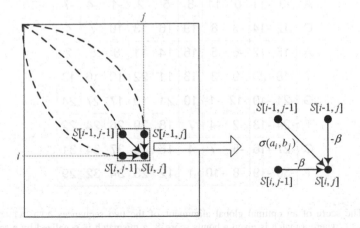

Fig. 3.4 There are three ways entering the grid point (i,j).

Algorithm GLOBAL_ALIGNMENT_SCORE($A = a_1a_2\ldots a_m, B = b_1b_2\ldots b_n$)
begin
   $S[0,0] \leftarrow 0$
   for $j \leftarrow 1$ **to** n **do** $S[0,j] \leftarrow -\beta \times j$
   for $i \leftarrow 1$ **to** m **do**
      $S[i,0] \leftarrow -\beta \times i$
      for $j \leftarrow 1$ **to** n **do**
$$S[i,j] \leftarrow \max \begin{cases} S[i-1,j] - \beta \\ S[i,j-1] - \beta \\ S[i-1,j-1] + \sigma(a_i,b_j) \end{cases}$$
   Output $S[m,n]$ as the score of an optimal alignment.
end

Fig. 3.5 Computation of the score of an optimal global alignment.

maximum score from $(5,2)$ is $-6-3 = -9$. Taking the maximum of them, we have $S[5,3] = 5$. Once the table has been computed, the value in the rightmost cell of the last row, i.e., $S[10,9] = 29$, is the score of an optimal global alignment.

In Section 2.4.4, we have shown that if a backtracking information is saved for each entry while we compute the dynamic-programming matrix, an optimal solution can be derived following the backtracking pointers. Here we show that even if we don't save those backtracking pointers, we can still reconstruct an optimal so-

		G	T	A	C	G	T	C	G	G
	0	-3	-6	-9	-12	-15	-18	-21	-24	-27
A	-3	-5	-8	2	-1	-4	-7	-10	-13	-16
T	-6	-8	3	0	-3	-6	4	1	-2	-5
A	-9	-11	0	11	8	5	2	-1	-4	-7
C	-12	-14	-3	8	19	16	13	10	7	4
A	-15	-17	-6	5	16	14	11	8	5	2
T	-18	-20	-9	2	13	11	22	19	16	13
G	-21	-10	-12	-1	10	21	19	17	27	24
T	-24	-13	-2	-4	7	18	29	26	24	22
C	-27	-16	-5	-7	4	15	26	37	34	31
T	-30	-19	-8	-10	1	12	23	34	32	29

Fig. 3.6 The score of an optimal global alignment of the two sequences ATACATGTCT and GTACGTCGG, where a match is given a bonus score 8, a mismatch is penalized by a score -5, and the gap penalty for each gap symbol is -3.

lution by examining the values of an entry's possible contributors. Figure 3.7 lists the pseudo-code for delivering an optimal global alignment, where an initial call GLOBAL_ALIGNMENT_OUTPUT(A, B, S, m, n) is made to deliver an optimal global alignment. Specifically, we trace back the dynamic-programming matrix from the entry (m, n) recursively according to the following rules. Let (i, j) be the entry under consideration. If $i = 0$ or $j = 0$, we simply output all the insertion pairs or deletion pairs in these boundary conditions. Otherwise, consider the following three cases. If $S[i, j] = S[i - 1, j - 1] + \sigma(a_i, b_j)$, we make a diagonal move and output a substitution pair $\begin{pmatrix} a_i \\ b_j \end{pmatrix}$. If $S[i, j] = S[i - 1, j] - \beta$, then we make a vertical move and output a deletion pair $\begin{pmatrix} a_i \\ - \end{pmatrix}$. Otherwise, it must be the case where $S[i, j] = S[i, j - 1] - \beta$. We simply make a horizontal move and output an insertion pair $\begin{pmatrix} - \\ b_j \end{pmatrix}$. Algorithm GLOBAL_ALIGNMENT_OUTPUT takes $O(m + n)$ time in total since each recursive call reduces i and/or j by one. The total space complexity is $O(mn)$ since the size of the dynamic-programming matrix is $O(mn)$. In Section 3.6, we shall show that an optimal global alignment can be recovered even if we don't save the whole matrix.

Figure 3.8 delivers an optimal global alignment by backtracking from the right-most cell of the last row of the dynamic-programming matrix computed in Figure 3.6. We start from the entry $(10, 9)$ where $S[10, 9] = 29$. We have a tie there because both $S[10, 8] - 3$ and $S[9, 8] - 5$ equal to 29. In this illustration, the horizontal move to the entry $(10, 8)$ is chosen. Interested readers are encouraged to try the

---

Algorithm GLOBAL_ALIGNMENT_OUTPUT($A = a_1 a_2 \ldots a_m$, $B = b_1 b_2 \ldots b_n$, S, i, j)
begin
   if $i = 0$ or $j = 0$ **then**
      if $i > 0$ **then for** $i' \leftarrow 1$ **to** i **do** print $\begin{pmatrix} a_{i'} \\ - \end{pmatrix}$
      if $j > 0$ **then for** $j' \leftarrow 1$ **to** j **do** print $\begin{pmatrix} - \\ b_{j'} \end{pmatrix}$
      return
   if $S[i, j] = S[i - 1, j - 1] + \sigma(a_i, b_j)$ **then**
      GLOBAL_ALIGNMENT_OUTPUT(A, B, S, $i - 1$, $j - 1$)
      print $\begin{pmatrix} a_i \\ b_j \end{pmatrix}$
   else if $S[i, j] = S[i - 1, j] - \beta$ **then**
      GLOBAL_ALIGNMENT_OUTPUT(A, B, S, $i - 1$, j)
      print $\begin{pmatrix} a_i \\ - \end{pmatrix}$
   else
      GLOBAL_ALIGNMENT_OUTPUT(A, B, S, i, $j - 1$)
      print $\begin{pmatrix} - \\ b_j \end{pmatrix}$
end

Fig. 3.7 Backtracking procedure for delivering an optimal global alignment.

		G	T	A	C	G	T	C	G	G
	0	-3	-6	-9	-12	-15	-18	-21	-24	-27
A	-3	-5	-8	2	-1	-4	-7	-10	-13	-16
T	-6	-8	-3	0	-3	-6	4	1	-2	-5
A	-9	-11	0	11	8	5	2	-1	-4	-7
C	-12	-14	-3	8	19	16	13	10	7	4
A	-15	-17	-6	5	16	14	11	8	5	2
T	-18	-20	-9	2	13	11	22	19	16	13
G	-21	-10	-12	-1	10	21	19	17	27	24
T	-24	-13	-2	-4	7	18	29	26	24	22
C	-27	-16	-5	-7	4	15	26	37	34	31
T	-30	-19	-8	-10	1	12	23	34	32	29

```
A T A C A T G T C T -
G T A C - - G T C G G
-5 +8 +8 +8 -3 -3 +8 +8 +8 -5 -3 = | 29 |
```

Fig. 3.8 Computation of an optimal global alignment of sequences ATACATGTCT and GTACGTCGG, where a match is given a bonus score 8, a mismatch is penalized by a score -5, and the gap penalty for each gap symbol is -3.

diagonal move to the entry $(9, 8)$ for an alternative optimal global alignment, which is actually chosen by GLOBAL_ALIGNMENT_OUTPUT. Continue this process until the entry $(0, 0)$ is reached. The shaded area depicts the backtracking path whose corresponding alignment is given on the right-hand side of the figure.

It should be noted that during the backtracking procedure, we derive the aligned pairs in a reverse order of the alignment. That's why we make a recursive call before actually printing out the pair in Figure 3.7. Another approach is to compute the dynamic-programming matrix backward from the rightmost cell of the last row to the leftmost cell of the first row. Then when we trace back from the leftmost cell of the first row toward the rightmost cell of the last row, the aligned pairs are derived in the same order as in the alignment. This approach could avoid the overhead of reversing an alignment.

3.4 Local Alignment

In many applications, a global (i.e., end-to-end) alignment of the two given sequences is inappropriate; instead, a local alignment (i.e., involving only a part of each sequence) is desired. In other words, one seeks a high-scoring local path that need not terminate at the corners of the dynamic-programming matrix.

Let $S[i, j]$ denote the score of the highest-scoring local path ending at (i, j) between $a_1 a_2 \ldots a_i$, and $b_1 b_2 \ldots b_j$. $S[i, j]$ can be computed as follows.

$$S[i,j] = \max \begin{cases} 0, \\ S[i-1,j] - \beta, \\ S[i,j-1] - \beta, \\ S[i-1,j-1] + \sigma(a_i, b_j). \end{cases}$$

The recurrence is quite similar to that for global alignment except the first entry "zero." For local alignment, we are not required to start from the source $(0,0)$. Therefore, if the scores of all possible paths ending at the current position are all negative, they are reset to zero. The largest value of $S[i,j]$ is the score of the best local alignment between sequences A and B.

Figure 3.9 gives the pseudo-code for computing the score of an optimal local alignment. Whenever there is a tie, any one of them will work. Since there are $O(mn)$ entries and the time spent for each entry is $O(1)$, the total running time of algorithm LOCAL_ALIGNMENT_SCORE is $O(mn)$.

Now let us use an example to illustrate the tabular computation. Figure 3.10 computes the score of an optimal local alignment of the two sequences ATACATGTCT and GTACGTCGG, where a match is given a bonus score 8, a mismatch is penalized by a score -5, and the gap penalty for each gap symbol is -3. The first row and column of the table are initialized with zero's. Other entries are computed in order. Take the entry $(5,5)$ for example. Upon computing the value of this entry, the following values are ready: $S[4,4] = 24$, $S[4,5] = 21$, and $S[5,4] = 21$. Since the edge weight of $(4,4) \to (5,5)$ is -5 (a mismatch), the maximum score from $(4,4)$ to $(5,5)$ is $24 - 5 = 19$. The maximum score from $(4,5)$ is $21 - 3 = 18$, and the maximum score from $(5,4)$ is $21 - 3 = 18$. Taking the maximum of them, we have $S[5,5] = 19$. Once

---

Algorithm LOCAL_ALIGNMENT_SCORE($A = a_1 a_2 \ldots a_m$, $B = b_1 b_2 \ldots b_n$)
begin
  $S[0,0] \leftarrow 0$
  $Best \leftarrow 0$
  $End_i \leftarrow 0$
  $End_j \leftarrow 0$
  for $j \leftarrow 1$ **to** n **do** $S[0,j] \leftarrow 0$
  for $i \leftarrow 1$ **to** m **do**
    $S[i,0] \leftarrow 0$
    for $j \leftarrow 1$ **to** n **do**

$$S[i,j] \leftarrow \max \begin{cases} 0 \\ S[i-1,j] - \beta \\ S[i,j-1] - \beta \\ S[i-1,j-1] + \sigma(a_i, b_j) \end{cases}$$

    if $S[i,j] > Best$ **then**
      $Best \leftarrow S[i,j]$
      $End_i \leftarrow i$
      $End_j \leftarrow j$
  Output $Best$ as the score of an optimal local alignment.
end

Fig. 3.9 Computation of the score of an optimal local alignment.

		G	T	A	C	G	T	C	G	G
	0	0	0	0	0	0	0	0	0	0
A	0	0	0	8	5	2	0	0	0	0
T	0	0	8	5	3	0	10	7	4	1
A	0	0	5	16	13	10	7	5	2	0
C	0	0	2	13	24	21	18	15	12	9
A	0	0	0	10	21	19	16	13	10	7
T	0	0	8	7	18	16	27	24	21	18
G	0	8	5	3	15	26	24	22	32	29
T	0	5	16	13	12	23	34	31	29	27
C	0	2	13	11	21	20	31	(42)	39	36
T	0	0	10	8	18	17	28	39	37	34

Fig. 3.10 Computation of the score of an optimal local alignment of the sequences ATACATGTCT and GTACGTCGG.

the table has been computed, the maximum value, *i.e.*, $S[9,7] = 42$, is the score of an optimal local alignment.

Figure 3.11 lists the pseudo-code for delivering an optimal local alignment, where an initial call LOCAL_ALIGNMENT_OUTPUT(A, B, S, End_i, End_j) is made to deliver an optimal local alignment. Specifically, we trace back the dynamic-programming matrix from the maximum-score entry (End_i, End_j) recursively according to the following rules. Let (i, j) be the entry under consideration. If $S[i, j] = 0$, we have reached the beginning of the optimal local alignment. Otherwise, consider the following three cases. If $S[i, j] = S[i-1, j-1] + \sigma(a_i, b_j)$, we make a diagonal move and output a substitution pair $\begin{pmatrix} a_i \\ b_j \end{pmatrix}$. If $S[i, j] = S[i-1, j] - \beta$, then we make a vertical move and output a deletion pair $\begin{pmatrix} a_i \\ - \end{pmatrix}$. Otherwise, it must be the case where $S[i, j] = S[i, j-1] - \beta$. We simply make a horizontal move and output an insertion pair $\begin{pmatrix} - \\ b_j \end{pmatrix}$. Algorithm LOCAL_ALIGNMENT_OUTPUT takes $O(m+n)$ time in total since each recursive call reduces i and/or j by one. The space complexity is $O(mn)$ since the size of the dynamic-programming matrix is $O(mn)$. In Section 3.6, we shall show that an optimal local alignment can be recovered even if we don't save the whole matrix.

Figure 3.12 delivers an optimal local alignment by backtracking from the maximum scoring entry of the dynamic-programming matrix computed in Figure 3.6.

Algorithm LOCAL_ALIGNMENT_OUTPUT($A = a_1a_2 \ldots a_m, B = b_1b_2 \ldots b_n, S, i, j$)
begin
   if $S[i,j] = 0$ **then return**
   if $S[i,j] = S[i-1,j-1] + \sigma(a_i,b_j)$ **then**
      LOCAL_ALIGNMENT_OUTPUT($A, B, S, i-1, j-1$)
      print $\begin{pmatrix} a_i \\ b_j \end{pmatrix}$
   else if $S[i,j] = S[i-1,j] - \beta$ **then**
      LOCAL_ALIGNMENT_OUTPUT($A, B, S, i-1, j$)
      print $\begin{pmatrix} a_i \\ - \end{pmatrix}$
   else
      LOCAL_ALIGNMENT_OUTPUT($A, B, S, i, j-1$)
      print $\begin{pmatrix} - \\ b_j \end{pmatrix}$
end

Fig. 3.11 Computation of an optimal local alignment.

We start from the entry $(9,7)$ where $S[9,7] = 42$. Since $S[8,6] + 8 = 34 + 8 = 42 = S[9,7]$, we make a diagonal move back to the entry $(8,6)$. Continue this process until an entry with zero value is reached. The shaded area depicts the backtracking path whose corresponding alignment is given on the right-hand side of the figure.

Further complications arise when one seeks k best alignments, where $k > 1$. For computing an arbitrary number of non-intersecting and high-scoring local alignments, Waterman and Eggert [198] developed a very time-efficient method. It records those high-scoring candidate regions of the dynamic-programming matrix in the first pass. Each time a best alignment is reported, it recomputes only those entries in the affected area rather than recompute the whole matrix. Its linear-space implementation was developed by Huang and Miller [92].

On the other hand, to attain greater speed, the strategy of building alignments from alignment fragments is often used. For example, one could specify some fragment length w and work with fragments consisting of a segment of length at least w that occurs exactly or approximately in both sequences. In general, algorithms that optimize the score over alignments constructed from fragments can run faster than algorithms that optimize over all possible alignments. Moreover, alignments constructed from fragments have been very successful in initial filtering criteria within programs that search a sequence database for matches to a query sequence. Database sequences whose alignment score with the query sequence falls below a threshold are ignored, and the remaining sequences are subjected to a slower but higher-resolution alignment process. The high-resolution process can be made more efficient by restricting the search to a "neighborhood" of the alignment-from-fragments. Chapter 4 will introduce four such homology search programs: FASTA, BLAST, BLAT, and PatternHunter.

		G	T	A	C	G	T	C	G	G
	0	0	0	0	0	0	0	0	0	0
A	0	0	0	8	5	2	0	0	0	0
T	0	0	8	5	3	0	10	7	4	1
A	0	0	5	16	13	10	7	5	2	0
C	0	0	2	13	24	21	18	15	12	9
A	0	0	0	10	21	19	16	13	10	7
T	0	0	8	7	18	16	27	24	21	18
G	0	8	5	3	15	26	24	22	32	29
T	0	5	16	13	12	23	34	31	29	27
C	0	2	13	11	21	20	31	42	39	36
T	0	0	10	8	18	17	28	39	37	34

```
T A C A T G T C
T A C - - G T C
+8 +8 +8 -3 -3 +8 +8 +8 = 42
```

Fig. 3.12 Computation of an optimal local alignment of the two sequences ATACATGTCT and GTACGTCGG.

3.5 Various Scoring Schemes

In this section, we shall briefly discuss how to modify the dynamic programming methods to copy with three scoring schemes that are frequently used in biological sequence analysis.

3.5.1 Affine Gap Penalties

For aligning DNA and protein sequences, affine gap penalties are considered more appropriate than the simple scoring scheme discussed in the previous sections. "Affine" means that a gap of length k is penalized $\alpha + k \times \beta$, where α and β are both nonnegative constants. In other words, it costs α to open up a gap plus β for each symbol in the gap. Figure 3.13 computes the score of a global alignment of the two sequences ATACATGTCT and GTACGTCGG under affine gap penalties, where a match is given a bonus score 8, a mismatch is penalized by a score -5, and the penalty for a gap of length k is $-4 - k \times 3$.

In order to determine if a gap is newly opened, two more matrices are used to distinguish gap extensions from gap openings. Let $D(i, j)$ denote the score of an optimal alignment between $a_1 a_2 \ldots a_i$ and $b_1 b_2 \ldots b_j$ ending with a deletion. Let $I(i, j)$ denote the score of an optimal alignment between $a_1 a_2 \ldots a_i$ and $b_1 b_2 \ldots b_j$ ending with an insertion. Let $S(i, j)$ denote the score of an optimal alignment between $a_1 a_2 \ldots a_i$ and $b_1 b_2 \ldots b_j$.

By definition, $D(i, j)$ can be derived as follows.

$$-4 \qquad\qquad\qquad -4$$
$$\downarrow \qquad\qquad\qquad\qquad \downarrow$$

```
A  T  A  C | A  T | G  T  C  T | - |
G  T  A  C | -  - | G  T  C  G | G |
-5 +8 +8 +8 -3 -3 +8 +8 +8 -5 -3 = 29
```

$$\boxed{29\text{-}4\text{-}4=21}$$

Fig. 3.13 The score of a global alignment of the two sequences ATACATGTCT and GTACGTCGG under affine gap penalties.

$$
\begin{aligned}
D(i,j) &= \max_{0 \le i' \le i-1} \{S(i',j) - \alpha - (i-i') \times \beta\} \\
&= \max\{ \max_{0 \le i' \le i-2} \{S(i',j) - \alpha - (i-i') \times \beta\}, S(i-1,j) - \alpha - \beta\} \\
&= \max\{ \max_{0 \le i' \le i-2} \{S(i',j) - \alpha - ((i-1)-i') \times \beta - \beta\}, S(i-1,j) - \alpha - \beta\} \\
&= \max\{D(i-1,j) - \beta, S(i-1,j) - \alpha - \beta\}.
\end{aligned}
$$

This recurrence can be explained in an alternative way. Recall that $D(i,j)$ denotes the score of an optimal alignment between $a_1 a_2 \ldots a_i$ and $b_1 b_2 \ldots b_j$ ending with a deletion. If such an alignment is an extension of the alignment ending at $(i-1,j)$ with a deletion, then it costs only β for such a gap extension. Thus, in this case, $D(i,j) = D(i-1,j) - \beta$. Otherwise, it is a new deletion gap and an additional gap-opening penalty α is imposed. We have $D(i,j) = S(i-1,j) - \alpha - \beta$.

In a similar way, we derive $I(i,j)$ as follows.

$$
\begin{aligned}
I(i,j) &= \max_{0 \le j' \le j-1} \{S(i,j') - \alpha - (j-j') \times \beta\} \\
&= \max\{ \max_{0 \le j' \le j-2} \{S(i,j') - \alpha - (j-j') \times \beta\}, S(i,j-1) - \alpha - \beta\} \\
&= \max\{I(i,j-1) - \beta, S(i,j-1) - \alpha - \beta\}.
\end{aligned}
$$

Therefore, with proper initializations, $D(i,j)$, $I(i,j)$ and $S(i,j)$ can be computed by the following recurrences:

$$
D(i,j) = \max \begin{cases} D(i-1,j) - \beta, \\ S(i-1,j) - \alpha - \beta; \end{cases}
$$

$$
I(i,j) = \max \begin{cases} I(i,j-1) - \beta, \\ S(i,j-1) - \alpha - \beta; \end{cases}
$$

$$S(i,j) = \max \begin{cases} D(i,j), \\ I(i,j), \\ S(i-1,j-1) + \sigma(a_i,b_j). \end{cases}$$

3.5.2 Constant Gap Penalties

Now let us consider the constant gap penalties where each gap, regardless of its length, is charged with a nonnegative constant penalty α.

Let $D(i,j)$ denote the score of an optimal alignment between $a_1a_2\ldots a_i$ and $b_1b_2\ldots b_j$ ending with a deletion. Let $I(i,j)$ denote the score of an optimal alignment between $a_1a_2\ldots a_i$ and $b_1b_2\ldots b_j$ ending with an insertion. Let $S(i,j)$ denote the score of an optimal alignment between $a_1a_2\ldots a_i$ and $b_1b_2\ldots b_j$. With proper initializations, $D(i,j)$, $I(i,j)$ and $S(i,j)$ can be computed by the following recurrences. In fact, these recurrences can be easily derived from those for the affine gap penalties by setting β to zero. A gap penalty is imposed when the gap is just opened, and the extension is free of charge.

$$D(i,j) = \max \begin{cases} D(i-1,j), \\ S(i-1,j) - \alpha; \end{cases}$$

$$I(i,j) = \max \begin{cases} I(i,j-1), \\ S(i,j-1) - \alpha; \end{cases}$$

$$S(i,j) = \max \begin{cases} D(i,j), \\ I(i,j), \\ S(i-1,j-1) + \sigma(a_i,b_j). \end{cases}$$

3.5.3 Restricted Affine Gap Penalties

Another interesting scoring scheme is called the restricted affine gap penalties, in which a gap of length k is penalized by $\alpha + f(k) \times \beta$, where α and β are both nonnegative constants, and $f(k) = \min\{k,\ell\}$ for a given positive integer ℓ.

In order to deal with the free long gaps, two more matrices $D'(i,j)$ and $I'(i,j)$ are used to record the long gap penalties in advance. With proper initializations, $D(i,j)$, $D'(i,j)$, $I(i,j)$, $I'(i,j)$, and $S(i,j)$ can be computed by the following recurrences:

$$D(i,j) = \max \begin{cases} D(i-1,j) - \beta, \\ S(i-1,j) - \alpha - \beta; \end{cases}$$

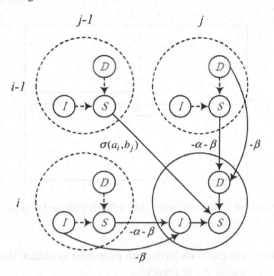

Fig. 3.14 There are seven ways entering the three grid points of an entry (i, j).

$$D'(i,j) = \max \begin{cases} D'(i-1,j), \\ S(i-1,j) - \alpha - \ell \times \beta; \end{cases}$$

$$I(i,j) = \max \begin{cases} I(i,j-1) - \beta, \\ S(i,j-1) - \alpha - \beta; \end{cases}$$

$$I'(i,j) = \max \begin{cases} I'(i,j-1), \\ S(i,j-1) - \alpha - \ell \times \beta; \end{cases}$$

$$S(i,j) = \max \begin{cases} D(i,j), \\ D'(i,j), \\ I(i,j), \\ I'(i,j), \\ S(i-1,j-1) + \sigma(a_i,b_j). \end{cases}$$

3.6 Space-Saving Strategies

Straightforward implementation of the dynamic-programming algorithms utilizes quadratic space to produce an optimal global or local alignment. For analysis of long DNA sequences, this space restriction is more crucial than the time constraint.

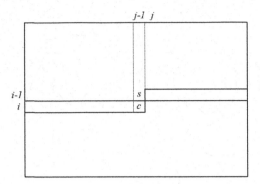

Fig. 3.15 Entry locations of S^- just before the entry value is evaluated at (i, j).

Because of this, different methods have been proposed to reduce the space used for aligning globally or locally two sequences.

We first describe a space-saving strategy proposed by Hirschberg in 1975 [91]. It uses only "linear space," *i.e.*, space proportional to the sum of the sequences' lengths. The original formulation was for the longest common subsequence problem that is discussed in Section 2.4.4. But the basic idea is quite robust and works readily for aligning globally two sequences with affine gap costs as shown by Myers and Miller in 1988 [150]. Remarkably, this space-saving strategy has the same time complexity as the original dynamic programming method presented in Section 3.3.

To introduce Hirschberg's approach, let us first review the original algorithm presented in Figure 3.5 for aligning two sequences of lengths m and n. It is apparent that the scores in row i of dynamic programming matrix S are calculated from those in row $i - 1$. Thus, after the scores in row i of S are calculated, the entries in row $i - 1$ of S will no longer be used and hence the space used for storing these entries can be recycled to calculate and store the entries in row $i + 1$. In other words, we can get by with space for two rows, since all that we ultimately want is the single entry $S[m, n]$ in the rightmost cell of the last row.

In fact, a single array S^- of size n, together with two extra variables, is adequate. $S^-[j]$ holds the most recently computed value for each $1 \le j \le n$, so that as soon as the value of the jth entry of S^- is computed, the old value at the entry is overwritten. There is a slight conflict in this strategy since we need the old value of an entry to compute a new value of the entry. To avoid this conflict, two additional variables, say s and c, are introduced to hold the new and old values of the entry, respectively. Figure 3.15 shows the locations of the scores kept in S^- and in variables s and c. When $S^-[j]$ is updated, $S^-[j']$ holds the score in the entry (i, j') in row i for each $j' < j$, and it holds the score in the entry $(i - 1, j')$ for any $j' \ge j$. Figure 3.16 gives the pseudo-code for computing the score of an optimal global alignment in linear space.

In the dynamic programming matrix S of aligning sequences $A = a_1 a_2 \ldots a_m$ and $B = b_1 b_2 \ldots b_n$, $S[i, j]$ denotes the optimal score of aligning $a_1 a_2 \ldots a_i$ and $b_1 b_2 \ldots b_j$

Algorithm FORWARD_SCORE($A = a_1 a_2 \ldots a_m$, $B = b_1 b_2 \ldots b_n$)
begin
    $S^-[0] \leftarrow 0$
    for $j \leftarrow 1$ **to** n **do** $S^-[j] \leftarrow S^-[j-1] - \beta$
    for $i \leftarrow 1$ **to** m **do**
        $s \leftarrow S^-[0]$
        $c \leftarrow S^-[0] - \beta$
        $S^-[0] \leftarrow c$
        for $j \leftarrow 1$ **to** n **do**
$$c \leftarrow \max \begin{cases} S^-[j] - \beta \\ c - \beta \\ s + \sigma(a_i, b_j) \end{cases}$$
        $s \leftarrow S^-[j]$
        $S^-[j] \leftarrow c$
    Output $S^-[n]$ as the score of an optimal alignment.
end

Fig. 3.16 Computation of the optimal score of aligning sequences of lengths m and n in linear space $O(n)$.

or, equivalently, the maximum score of a path from $(0,0)$ to the cell (i, j) in the alignment graph. By symmetry, the optimal score of aligning $a_{i+1} a_{i+2} \ldots a_m$ and $b_{j+1} b_{j+2} \ldots b_n$ or the maximum score of a path from (i, j) to (m, n) in the alignment graph can be calculated in linear space in a backward manner. Figure 3.17 gives the pseudo-code for computing the score of an optimal global alignment in a backward manner in linear space.

In what follows, we use $S^-[i, j]$ and $S^+[i, j]$ to denote the maximum score of a path from $(0,0)$ to (i, j) and that from (i, j) to (m, n) in the alignment graph, respectively. Without loss of generality, we assume that m is a power of 2. Obviously, for each j, $S^-[m/2, j] + S^+[m/2, j]$ is the maximum score of a path from $(0,0)$ to (m, n) through $(m/2, j)$ in the alignment graph. Choose j_{mid} such that

$$S^-[m/2, j_{mid}] + S^+[m/2, j_{mid}] = \max_{1 \leq j \leq n} S^-[m/2, j] + S^+[m/2, j].$$

Then, $S^-[m/2, j_{mid}] + S^+[m/2, j_{mid}]$ is the optimal alignment score of A and B and there is a path having such a score from $(0,0)$ to (m, n) through $(m/2, j_{mid})$ in the alignment graph.

Hirschberg's linear-space approach is first to compute $S^-[m/2, j]$ for $1 \leq j \leq n$ by a forward pass, stopping at row $m/2$ and to compute $S^+[m/2, j]$ for $1 \leq j \leq n$ by a backward pass and then to find j_{mid}. After j_{mid} is found, recursively compute an optimal path from $(0,0)$ to $(m/2, j_{mid})$ and an optimal path from $(m/2, j_{mid})$ to (m, n).

As the problem is partitioned further, there is a need to have an algorithm that is capable of delivering an optimal path for any specified two ends. In Figure 3.18, algorithm LINEAR_ALIGN is a recursive procedure that delivers a maximum-scoring path from (i_1, j_1) to (i_2, j_2). To deliver the whole optimal alignment, the two ends are initially specified as $(0,0)$ and (m, n).

Now let us analyze the time and space taken by Hirschberg's approach. Using the algorithms given in Figures 3.16 and 3.17, both the forward and backward pass take $O(nm/2)$-time and $O(n)$-spaces. Hence, it takes $O(mn)$-time and $O(n)$-spaces to find j_{mid}. Set $T = mn$ and call it the size of the problem of aligning A and B. At each recursive step, a problem is divided into two subproblems. However, regardless of where the optimal path crosses the middle row $m/2$, the total size of the two resulting subproblems is exactly half the size of the problem that we have at the recursive step (see Figure 3.19). It follows that the total size of all problems, at all levels of recursion, is at most $T + T/2 + T/4 + \cdots = 2T$. Because computation time is directly proportional to the problem size, Hirschberg's approach will deliver an optimal alignment using $O(2T) = O(T)$ time. In other words, it yields an $O(mn)$-time, $O(n)$-space global alignment algorithm.

Hirschberg's original method, and the above discussion, apply to the case where the penalty for a gap is merely proportional to the gap's length, i.e., $k \times \beta$ for a k-symbol gap. For applications in molecular biology, one wants penalties of the form $\alpha + k \times \beta$, i.e., each gap is assessed an additional "gap-open" penalty α. Actually, one can be slightly more general and substitute residue-dependent penalties for β. In Section 3.5.1, we have shown that the relevant alignment graph is more complicated. Now at each grid point (i, j) there are three nodes, denoted $(i, j)_S$, $(i, j)_D$, and $(i, j)_I$, and generally seven entering edges, as pictured in Figure 3.14. The alignment problem is to compute a highest-score path from $(0, 0)_S$ to $(m, n)_S$. Fortunately, Hirschberg's strategy extends readily to this more general class of alignment scores [150]. In essence, the main additional complication is that for each defining corner of a subproblem, we need to specify one of the grid point's three nodes.

Another issue is how to deliver an optimal *local* alignment in linear space. Recall that in the local alignment problem, one seeks a highest-scoring alignment where the end nodes can be arbitrary, *i.e.*, they are not restricted to $(0, 0)_S$ and $(m, n)_S$. In fact, it can be reduced to a global alignment problem by performing a linear-space

---

Algorithm BACKWARD_SCORE($A = a_1 a_2 \ldots a_m$, $B = b_1 b_2 \ldots b_n$)
begin
   $S^+[n] \leftarrow 0$
   for $j \leftarrow n - 1$ **down to** 0 **do** $S^+[j] \leftarrow S^+[j+1] - \beta$
   for $i \leftarrow m - 1$ **down to** 0 **do**
      $s \leftarrow S^+[n]$
      $c \leftarrow S^+[n] - \beta$
      $S^+[n] \leftarrow c$
      for $j \leftarrow n - 1$ **down to** 0 **do**
         $c \leftarrow \max \begin{cases} S^+[j] - \beta \\ c - \beta \\ s + \sigma(a_{i+1}, b_{j+1}) \end{cases}$
         $s \leftarrow S^+[j]$
         $S^+[j] \leftarrow c$
   Output $S^+[0]$ as the score of an optimal alignment.
end

Fig. 3.17 Backward computation of the score of an optimal global alignment in linear space.

score-only pass over the dynamic-programming matrix to locate the first and last nodes of an optimal local alignment, then delivering a global alignment between these two nodes by applying Hirschberg's approach.

Algorithm LINEAR_ALIGN($A = a_1 a_2 \ldots a_m$, $B = b_1 b_2 \ldots b_n$, i_1, j_1, i_2, j_2)
begin
    if $i_1 + 1 \leq i_2$ or $j_1 + 1 \leq j_2$ **then**
        Output the aligned pairs for the maximum-score path from (i_1, j_1) to (i_2, j_2)
    else
        $i_{mid} \leftarrow \lfloor (i_1 + i_2)/2 \rfloor$
        // Find the maximum scores from (i_1, j_1)
        $S^-[j_1] \leftarrow 0$
        for $j \leftarrow j_1 + 1$ **to** j_2 **do** $S^-[j] \leftarrow S^-[j-1] - \beta$
        for $i \leftarrow i_1 + 1$ **to** i_{mid} **do**
            $s \leftarrow S^-[j_1]$
            $c \leftarrow S^-[j_1] - \beta$
            $S^-[j_1] \leftarrow c$
            for $j \leftarrow j_1 + 1$ **to** j_2 **do**
                $c \leftarrow \max \begin{cases} S^-[j] - \beta \\ c - \beta \\ s + \sigma(a_i, b_j) \end{cases}$
                $s \leftarrow S^-[j]$
                $S^-[j] \leftarrow c$
        // Find the maximum scores to (i_2, j_2)
        $S^+[j_2] \leftarrow 0$
        for $j \leftarrow j_2 - 1$ **down to** j_1 **do** $S^+[j] \leftarrow S^+[j+1] - \beta$
        for $i \leftarrow i_2 - 1$ **down to** i_{mid} **do**
            $s \leftarrow S^+[j_2]$
            $c \leftarrow S^+[j_2] - \beta$
            $S^+[j_2] \leftarrow c$
            for $j \leftarrow j_2 - 1$ **down to** j_1 **do**
                $c \leftarrow \max \begin{cases} S^+[j] - \beta \\ c - \beta \\ s + \sigma(a_{i+1}, b_{j+1}) \end{cases}$
                $s \leftarrow S^+[j]$
                $S^+[j] \leftarrow c$
        // Find where maximum-score path crosses row i_{mid}
        $j_{mid} \leftarrow$ value $j \in [j_1, j_2]$ that maximizes $S^-[j] + S^+[j]$
        LINEAR_ALIGN($A, B, i_1, j_1, i_{mid}, j_{mid}$)
        LINEAR_ALIGN($A, B, i_{mid}, j_{mid}, i_2, j_2$)
end

Fig. 3.18 Computation of an optimal global alignment in linear space.

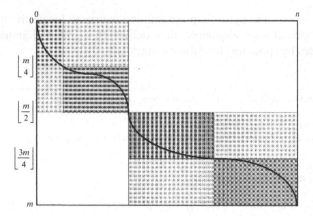

Fig. 3.19 Hirschberg's linear-space approach.

3.7 Other Advanced Topics

In this section, we discuss several advanced topics such as constrained sequence alignment, similar sequence alignment, suboptimal alignment, and robustness measurement.

3.7.1 Constrained Sequence Alignment

Rigorous sequence alignment algorithms compare each residue of one sequence to every residue of the other. This requires computational time proportional to the product of the lengths of the given sequences. Biologically relevant sequence alignments, however, usually extend from the beginning of both sequences to the end of both sequences, and thus the rigorous approach is unnecessarily time consuming; significant sequence similarities are rarely found by aligning the end of one sequence with the beginning of the other.

As a result of the biological constraint, it is frequently possible to calculate an optimal alignment between two sequences by considering only those residues that are within a diagonal band in which each row has only w cells. With sequences $A = a_1 a_2 \ldots a_m$ and $B = b_1 b_2 \ldots b_n$, one can specify constants $\ell \leq u$ such that aligning a_i with b_j is permitted only if $\ell \leq j - i \leq u$. For example, it rarely takes a dozen insertions or deletions to align any two members of the globin superfamily; thus, an optimal alignment of two globin sequences can be calculated in $O(nw)$ time that is identical to the rigorous alignment that requires $O(nm)$ time.

Alignment within a band is used in the final stage of the FASTA program for rapid searching of protein and DNA sequence databases (Pearson and Lipman, 1988; Pearson, 1990). For optimization in a band, the requirement to "start at the beginning, end at the end" is reflected in the $\ell \leq \min\{0, n-m\}$ and $u \geq \max\{0, n-m\}$ constraints. "Local" sequence alignments do not require that the beginning and end of the alignment correspond to the beginning and end of the sequence, *i.e.*, the aligned sequences can be arbitrary substrings of the given sequences, A and B; they simply require that the alignment have the highest similarity score. For a "local" alignment in a band, it is natural to relax the requirement to $\ell \leq u$. Algorithms for computing an optimal local alignment can utilize a global alignment procedure to perform subcomputations: once locally optimal substrings A' of A and B' of B are found, which can be done by any of several available methods, a global alignment procedure is called to align A' and B'. Appropriate values of ℓ' and u' for the global problem are inferred from the ℓ and u of the local problems. In other situations, a method to find unconstrained local alignments, *i.e.*, without band limits, might determine appropriate values of ℓ and u before invoking a global alignment procedure within a band.

Although the application of rigorous alignment algorithms to long sequences can be quite time-consuming, it is often the space requirement that is limiting in practice. Hirschberg's approach, introduced in Section 3.6, can be easily modified to find a solution locating in a band. Unfortunately, the resulting time required to produce the alignment can exceed that of the score-only calculation by a substantial factor. If T denotes the number of entries in the band of the dynamic programming matrix, then $T = O(nw)$. Producing an alignment involves computing as many as $T \times \log_2 n$ entries (including recomputations of entries evaluated at earlier steps). Thus, the time to deliver an alignment exceeds that for computing its score in a band by a log factor.

To avoid the log factor, we need a new way to subdivide the problem that limits the subproblems to some fraction, $\alpha < 1$, of the band. Figure 3.20 illustrates the idea. The score-only backward pass is augmented so that at each point it computes the next place where an optimal path crosses the mid-diagonal, *i.e.*, diagonal $(\ell+u)/2$. Using only linear space, we can save this information at every point on the "current row" or on the mid-diagonal. When this pass is completed, we can use the retained information to find the sequence of points where an optimal solution crosses the mid-diagonal, which splits the problem into some number of subproblems. The total area of these subproblems is no more than half of the original area for a narrow band with widely spaced crossing points; in other cases it is even less.

It should be noted that this band-aligning algorithm could be considered as a generalization of Hirschberg's approach by rotating the matrix partition line. The idea of partition line rotation has been exploited in devising parallel sequence comparison algorithms. Nevertheless, the dividing technique proposed in this section, which produces more than two subproblems, reveals a new paradigm for space-saving strategies.

Another extension is to consider the situations where the ith entry of the first sequence can be aligned to the jth entry of the second sequence only if $L[i] \leq j \leq U[i]$, for given left and right bounds L and U. As in the band alignment problem, we can apply the idea of defining a midpoint *partition line* that bisects the region

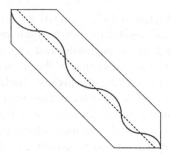

Fig. 3.20 Dividing a band by its middle diagonal.

into two nearly equal parts. Here we introduce a more general approach that can be easily utilized by other relevant problems.

Given a narrow region R with two boundary lines L and U, we can proceed as follows. We assume that L and U are non-decreasing since if, e.g., $L[i]$ were larger than $L[i+1]$, we could set $L[i+1]$ to equal $L[i]$ without affecting the set of constrained alignments. Enclose as many rows as possible from the top of the region in an upright rectangle, subject to the condition that the rectangle's area at most doubles the area of its intersection with R. Then starting with the first row of R not in the rectangle, we cover additional rows of R with a second such rectangle, and so on.

A score-only backward pass is made over R, computing S^+. Values of S^+ are retained for the top line in every rectangle (the top rectangle can be skipped). It can be shown that the total length of these pieces cannot exceed three times the total number of columns, as required for a linear space bound. Next, perform a score-only forward pass, stopping at the last row in the first rectangle. A sweep along the boundary between the first and second rectangles locates a crossing edge on an optimal path through R. That is, we can find a point p on the last row of the first rectangle and a point q on the first row of the second rectangle such that there is a vertical or diagonal edge e from p to q, and e is on an optimal path. Such an optimal path can be found by applying Hirschberg's strategy to R's intersection with the first rectangle (omitting columns following p) and recursively computing a path from q through the remainder of R. This process inspects a grid point at most once during the backward pass, once in a forward pass computing p and q, and an average of four times for applying Hirschberg's method to R's intersection with a rectangle.

3.7.2 Similar Sequence Alignment

If two sequences are very similar, more efficient algorithms can be devised to deliver an optimal alignment for them. In this case, we know that the maximum-scoring path in the alignment graph will not get too far away from the diagonal of the source

$(0,0)$. One way is to draw a constrained region to restrict the diversion and run the constrained alignment algorithm introduced in Section 3.7.1. Another approach is to grow the path greedily until the destination (m,n) is reached. For this approach, instead of working with the maximization of the alignment score, we look for the minimum-cost set of single-nucleotide changes (i.e., insertions, deletions, or substitutions) that will convert one sequence to the other. Any match costs zero, which allows us to have a free advance. As for penalty, we have to pay a certain amount of cost to get across a mismatch or a gap symbol.

Now we briefly describe the approach based on the diagonalwise monotonicity of the cost tables. The following cost function is employed. Each match costs 0, each mismatch costs 1, and a gap of length k is penalized at the cost of $k+1$. Adding 1 to a gap's length to derive its cost decreases the likelihood of generating gaps that are separated by only a few paired nucleotides. The edit graph for sequences A and B is a directed graph with a vertex at each integer grid point (x,y), $0 \le x \le m$ and $0 \le y \le n$. Let $I(z,c)$ denote the x value of the farthest point in diagonal z $(=y-x)$ that can be reached from the source (i.e., grid point $(0,0)$) with cost c and that is *free* to open an insertion gap. That is, the grid point can be (1) reached by a path of cost c that ends with an insertion, or (2) reached by any path of cost $c-1$ and the gap-open penalty of 1 can be "paid in advance." (The more traditional definition, which considers only case (1), results in the storage of more vectors.) Let $D(z,c)$ denote the x value of the farthest point in diagonal z that can be reached from the source with cost c and is *free* to open a deletion gap. Let $S(z,c)$ denote the x value of the farthest point in diagonal z that can be reached from the source with cost c. With proper initializations, these vectors can be computed by the following recurrence relations:

$$I(z,c) = \max\{I(z-1,c-1), S(z,c-1)\},$$

$$D(z,c) = \max\{D(z+1,c-1)+1, S(z,c-1)\},$$

$$S(z,c) = snake(z, \max\{S(z,c-1)+1, I(z,c), D(z,c)\}),$$

where $snake(z,x) = \max\{x, \max\{t : a_x \ldots a_{t-1} = b_{x+z} \ldots b_{t-1+z}\}\}$.

Since vectors at cost c depend only on those at costs c and $c-1$, it is straightforward to derive a dynamic-programming algorithm from the above recurrence relations.

3.7.3 Suboptimal Alignment

Molecular biology is rapidly becoming a data-rich science with extensive computational needs. More and more computer scientists are working together on developing efficient software tools for molecular biologists. One major area of potential interaction between computer scientists and molecular biologists arises from the need for analyzing biological information. In particular, optimal alignments mentioned in previous sections have been used to reveal similarities among biological sequences, to study gene regulation, and even to infer evolutionary trees.

However, biologically significant alignments are not necessarily mathematically optimized. It has been shown that sometimes the neighborhood of an optimal alignment reveals additional interesting biological features. Besides, the most strongly conserved regions can be effectively located by inspecting the range of variation of suboptimal alignments. Although rigorous statistical analysis for the mean and variance of optimal global alignment scores is not yet available, suboptimal alignments have been successfully used to informally estimate the significance of an optimal alignment.

For most applications, it is impractical to enumerate all suboptimal alignments since the number could be enormous. Therefore, a more compact representation of all suboptimal alignments is indispensable. A 0-1 matrix can be used to indicate if a pair of positions is in some suboptimal alignment or not. However, this approach misses some connectivity information among those pairs of positions. An alternative is to use a set of "canonical" suboptimal alignments to represent all suboptimal alignments. The kernel of that representation is a minimal directed acyclic graph (DAG) containing all suboptimal alignments.

Suppose we are given a threshold score that does not exceed the optimal alignment score. An alignment is suboptimal if its score is at least as large as the threshold score. Here we briefly describe a linear-space method that finds all edges that are contained in at least one path whose score exceeds a given threshold τ. Again, a recursive subproblem will consist of applying the alignment algorithm over a rectangular portion of the original dynamic-programming matrix, but now it is necessary that we continue to work with values S^- and S^+ that are defined relative to the original problem. To accomplish this, each problem to be solved is defined by specifying values of S^- for nodes on the upper and left borders of the defining rectangle, and values of S^+ for the lower and right borders.

To divide a problem of this form, a forward pass propagates values of S^- to nodes in the middle row and the middle column, and a backward pass propagates values S^+ to those nodes. This information allows us to determine all edges starting in the middle row or middle column that are contained in a path of score at least τ. The data determining any one of the four subproblems, $i.e.$, the arrays of S values on its borders, is then at most half the size of the set of data defining the parent problem. The maximum total space requirement is realized when recursion reaches a directly solvable problem where there is only the leftmost cell of the first row of the original grid left; at that time there are essentially $2(m+n)$ S-values saved for borders of the original problem, $m+n$ values on the middle row and column of the original problem, $(m+n)/2$ values for the upper left subproblem, $(m+n)/4$ values for the upper-left-most subsubproblem, etc., giving a total of about $4(m+n)$ retained S-values.

3.7.4 Robustness Measurement

The utility of information about the reliability of different regions within an alignment is widely appreciated, see [192] for example. One approach to obtaining such information is to determine suboptimal alignments, *i.e.*, some or all alignments that come within a specified tolerance of the optimum score, as discussed in Section 3.7.3. However, the number of suboptimal alignments, or even alternative optimal alignments, can easily be so large as to preclude an exhaustive enumeration.

Sequence conservation has proved to be a reliable indicator of at least one class of regulatory elements. Specifically, regions of six or more consecutive nucleotides that identical across a range of mammalian sequences, called "phylogenetic footprints," frequently correspond to binding sits for sequence-specific nuclear proteins. It is also interesting to look for longer, imperfectly conserved (but stronger matching) regions, which may indicate other sorts of regulatory elements, such as a region that binds to a nuclear matrix or assumes some altered chromatin structure.

In the following, we briefly describe some interesting measurements of the robustness of each aligned pair of a pairwise alignment. The first method computes, for each position i of the first sequence, the lower and upper limits of the positions in the second sequence to which it can be aligned and still come within a specified tolerance of the optimum alignment score. Delimiting suboptimal alignments this way, rather than enumerating all of them, allows the computation to run in only a small constant factor more time than the computation of a single optimal alignment.

Another method determines, for each aligned pair of an optimal alignment, the amount by which the optimum score must be lowered before reaching an alignment not containing that pair. In other words, if the optimum alignment score is s and the aligned pair is assigned the robustness-measuring number r, then any alignment scoring strictly greater than $s - r$ aligns those two sequence positions, whereas some alignment of score $s - r$ does not align them. As a special case, this value

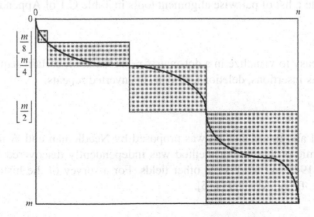

Fig. 3.21 The total number of the boundary entries in the active subproblems is $O(m+n)$.

tells whether the pair is in all optimal alignments (namely, the pair is in all optimal alignments if and only if its associated value is non-zero). These computations are performed using dynamic-programming methods that require only space proportional to the sum of the two sequence lengths. It has also been shown on how to efficiently handle the case where alignments are constrained so that each position, say position i, of the first sequence can be aligned only to positions on a certain range of the second sequence.

To deliver an optimal alignment, Hirschberg's approach applies forward and backward passes in the first nondegenerate rectangle along the optimal path being generated. Within a subproblem (i.e., rectangle) the scores of paths can be taken relative to the "start node" at the rectangle's upper left and the "end node" at the rightmost cell of the last row. This means that a subproblem is completely specified by giving the coordinates of those two nodes. In contrast, methods for the robustness measurement must maintain more information about each pending subproblem. Fortunately, it can be done in linear space by observing that the total number of the boundary entries of all pending subproblems of Hirschberg's approach is bounded by $O(m+n)$ (see Figure 3.21).

3.8 Bibliographic Notes and Further Reading

Sequence alignment is one of the most fundamental components of bioinformatics. For more references and applications, see the books by Sankoff and Kruskal [175], Waterman [197], Gusfield [85], Durbin et al. [61], Pevzner [165], Jones and Pevzner [98], and Deonier et al. [58], or recent survey papers by Batzoglou [23] and Notredame [154].

3.1

We compile a list of pairwise alignment tools in Table C.1 of Appendix C.

3.2

It is very easy to visualize in a dot-matrix representation certain sequence similarities such as insertions, deletions, repeats, or inverted repeats.

3.3

The global alignment method was proposed by Needleman and Wunsch [151]. Such a dynamic-programming method was independently discovered by Wagner and Fischer [194] and workers in other fields. For a survey of the history, see the book by Sankoff and Kruskal [175].

3.4

The local alignment method is so-called the *Smith-Waterman algorithm* [180]. In most applications of pairwise alignment, affine gap penalties are used [78].

3.5

Biologists need more general measurements of sequence relatedness than are typically considered by computer scientists. The most popular formulation in the computer science literature is the "longest common subsequence problem," which is equivalent to scoring alignments by simply counting the number of exact matches. For comparing protein sequences, it is important to reward alignment of residues that are similar in functions [70].

For both DNA and protein sequences, it is standard to penalize a long "gap," *i.e.*, a block of consecutive dashes, less than the sum of the penalties for the individual dashes in the gap. In reality, a gap would most likely represent a single insertion or deletion of a block of letters rather than multiple insertions or deletions of single letters [71]. This is usually accomplished by charging $\alpha + k \times \beta$ for a gap of length k. Thus the "gap-open penalty" α is assessed for every gap, regardless of length, and an additional "gap-extension penalty" β is charged for every dash in the gap. Such penalties are called *affine* gap penalties. Gotoh [78] showed how to efficiently compute optimal alignments under such a scoring scheme.

Even more general models for quantifying sequence relatedness have been proposed. For example, it is sometimes useful to let the penalty for adding a symbol to a gap depend on the position of the gap within the sequence [81], which is motivated by the observation that insertions in certain regions of a protein sequence can be much more likely than at other regions. Another generalization is to let the incremental gap score $\delta_i = c_{i+1} - c_i$, where a k-symbol gap scores c_k, be a monotone function of i, *e.g.*, $\delta_1 \geq \delta_2 \geq \cdots$ [139, 196].

Gotoh [79] proposed the piecewise linear gap penalties to allow long gaps in a resulting alignment. Huang and Chao [93] generalized the global alignment algorithms to compare sequences with intermittent similarities, an ordered list of similar regions separated by different regions.

Most alignment methods can be extended to deal with free end gaps in a straightforward way.

3.6

Readers can refer to [40, 41, 43] for more space-saving strategies.

3.7

Readers should also be aware that the hidden Markov models are a probabilistic approach to sequence comparison. They have been widely used in the bioinformatics community [61]. Given an observed sequence, the Viterbi algorithm computes the most probable state path. The forward algorithm computes the probability that a given observed sequence is generated by the model, whereas the backward algorithm computes the probability that a given observed symbol was generated by a given state. The book by Durbin et al. [61] is a terrific reference book for this paradigm.

Alignment of two genomic sequences poses problems not well addressed by earlier alignment programs.

PipMaker [178] is a software tool for comparing two long DNA sequences to identify conserved segments and for producing informative, high-resolution displays of the resulting alignments. It displays a percent identity plot (pip), which shows both the position in one sequence and the degree of similarity for each aligning segment between the two sequences in a compact and easily understandable form. The alignment program used by the PipMaker network server is called BLASTZ [177]. It is an independent implementation of the Gapped BLAST algorithm specifically designed for aligning two long genomic sequences. Several modifications have been made to BLASTZ to attain efficiency adequate for aligning entire mammalian genomes and to increase the sensitivity.

MUMmer [119] is a system for aligning entire genomes rapidly. The core of the MUMmer algorithm is a suffix tree data structure, which can be built and searched in linear time and which occupies only linear space. DisplayMUMs 1.0 graphically presents alignments of MUMs from a set of query sequences and a single reference sequence. Users can navigate MUM alignments to visually analyze coverage, tiling patterns, and discontinuities due to misassemblies or SNPs.

The analysis of genome rearrangements is another exciting field for whole genome comparison. It looks for a series of genome rearrangements that would transform one genome into another. It was pioneered by Dobzhansky and Sturtevant [60] in 1938. Recent milestone advances include the works by Bafna and Pevzner [19], Hannenhalli and Pevzner [86], and Pevzner and Tesler [166].

Chapter 4
Homology Search Tools

The alignment methods introduced in Chapter 3 are good for comparing two sequences accurately. However, they are not adequate for homology search against a large biological database such as GenBank. As of February 2008, there are approximately 85,759,586,764 bases in 82,853,685 sequence records in the traditional GenBank divisions. To search such kind of huge databases, faster methods are required for identifying the homology between the query sequence and the database sequence in a timely manner.

One common feature of homology search programs is the filtration idea, which uses exact matches or approximate matches between the query sequence and the database sequence as a basis to judge if the homology between the two sequences passes the desired threshold.

This chapter is divided into six sections. Section 4.1 describes how to implement the filtration idea for finding exact word matches between two sequences by using efficient data structures such as hash tables, suffix trees, and suffix arrays.

FASTA was the first popular homology search tool, and its file format is still widely used. Section 4.2 briefly describes a multi-step approach used by FASTA for finding local alignments.

BLAST is the most popular homology search tool now. Section 4.3 reviews the first version of BLAST, *Ungapped BLAST*, which generates ungapped alignments. It then reviews two major products of BLAST 2.0: Gapped BLAST and Position-Specific Iterated BLAST (PSI-BLAST). Gapped BLAST produces gapped alignments, yet it is able to run faster than the original one. PSI-BLAST can be used to find distant relatives of a protein based on the profiles derived from the multiple alignments of the highest scoring database sequence segments with the query segment in iterative Gapped BLAST searches.

Section 4.4 describes BLAT, short for "BLAST-like alignment tool." It is often used to search for the database sequences that are closely related to the query sequences such as producing mRNA/DNA alignments and comparing vertebrate sequences.

PatternHunter, introduced in Section 4.5, is more sensitive than BLAST when a hit contains the same number of matches. A novel idea in PatternHunter is the use of an optimized spaced seed. Furthermore, it has been demonstrated that using optimized multiple spaced seed will speed up the computation even more.

Finally, we conclude the chapter with the bibliographic notes in Section 4.6.

4.1 Finding Exact Word Matches

An exact word match is a run of identities between two sequences. In the following, we discuss how to find all short exact word matches, sometimes referred to as *hits*, between two sequences using efficient data structures such as hash tables, suffix trees, and suffix arrays.

Given two sequences $A = a_1 a_2 \ldots a_m$, and $B = b_1 b_2 \ldots b_n$, and a positive integer k, the *exact word match* problem is to find all occurrences of exact word matches of length k, referred to as k-mers between A and B. This is a classic algorithmic problem that has been investigated for decades. Here we describe three approaches for this problem.

4.1.1 Hash Tables

A hash table associates keys with numbers. It uses a hash function to transform a given key into a number, called hash, which is used as an index to look up or store the corresponding data. A method that uses a hash table for finding all exact word matches of length w between two DNA sequences A and B is described as follows.

Since a DNA sequence is a sequence of four letters A, C, G, and T, there are 4^w possible DNA w-mers. The following encoding scheme maps a DNA w-mer to an integer between 0 and $4^w - 1$. Let $C = c_1 c_2 \ldots c_w$ be a w-mer. The hash value of C is written $V(C)$ and its value is

$$V(C) = x_1 \times 4^{w-1} + x_2 \times 4^{w-2} + \cdots + x_w \times 4^0,$$

where $x_i = 0, 1, 2, 3$ if $c_i = $ A, C, G, T, respectively. For example, if C=GTCAT, then

$$V(C) = 2 \times 4^4 + 3 \times 4^3 + 1 \times 4^2 + 0 \times 4^1 + 3 \times 4^0 = 732.$$

In fact, we can use two bits to represent each nucleotide: A(00), C(01), G(10), and T(11). In this way, a DNA segment is transformed into a binary string by compressing four nucleotides into one byte. For C=GTCAT given above, we have

$$V(C) = 732 = 1011010011_2.$$

Initially, a hash table H of size 4^w is created. To find all exact word matches of length w between two sequences A and B, the following steps are executed. The first step is to hash sequence A into a table. All possible w-mers in A are calculated by

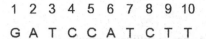

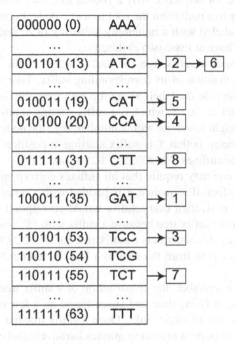

Fig. 4.1 A 3-mer hash table for GATCCATCTT.

sliding a window of size w from position 1 to position $m - w + 1$. For each word C, we compute $V(C)$ and insert C to the entry $H[V(C)]$. If there is more than one window word having the same hash value, a linked list or an array can be used to store them. Figure 4.1 depicts the process of constructing a hash table of word size 3 for GATCCATCTT.

Once a hash table for sequence A has been built, we can now scan sequence B by sliding a window of size w from position 1 to position $n - w + 1$. For each scan S, we compute $V(S)$ and find its corresponding exact word matches, if any, in A by looking up the entry $H[V[S]$ in the hash table H. All the exact word matches can be found in an order of their occurrences.

A hash table works well in practice for a moderate word size, say 12. However, it should be noted that for some larger word sizes, this approach might not be feasible. Suppose an exact word match of interest has 40 nucleotides. There are 4^{40} possible combinations. If we use an array to store all possible keys, then we would need 4^{40} entries to assign a different index to each combination, which would be far beyond the capacity of any modern computers. A more succinct indexing technique, such as suffix trees or suffix arrays, is required for this particular application.

4.1.2 Suffix Trees

A sequence $A = a_1 a_2 \ldots a_m$ has m suffixes, namely, $a_1 \ldots a_m, a_2 \ldots a_m, a_3 \ldots a_m, \ldots,$ and a_m. A suffix tree for sequence A is a rooted tree such that every suffix of A corresponds uniquely to a path from the root to a tree node. Furthermore, each edge of the suffix tree is labeled with a nonempty substring of A, and all internal nodes except the root must have at lease two children.

Figure 4.2 constructs a suffix tree for GATCCATCTT. The number of a node specifies the starting position of its corresponding suffix. Take the number "5" for example. If we concatenate the labels along the path from the root to the node with the number "5," we get CATCTT, which is a suffix starting at position 5. Notice that some internal node might associate with a number, *e.g.*, the node with number "10" in this figure. The reason is that T is suffix starting at position 10, yet it is also a prefix of another three suffixes TCCATCTT, TCTT, and TT.

For convenience, one may require that all suffixes correspond to paths from the root to the *leaves*. In fact, if sequence A is padded with a terminal symbol, say $, that does not appear in A, then every suffix would correspond to a path from the root to a leaf node in the suffix tree because a suffix with "$" will not be a prefix of any other suffix. Figure 4.3 constructs a suffix tree for GATCCATCTT$. Now every suffix corresponds to a path from the root to a leaf, including the suffix starting at position 10.

For a constant-size alphabet, the construction of a suffix tree for a sequence of length m can be done in $O(m)$ time and space based on a few other crucial observations, including the use of suffix links. Once a suffix tree has been built, we can answer several kinds of pattern matching queries iteratively and efficiently. Take the exact word match problem for example. Given are two sequences $A = a_1 a_2 \ldots a_m$,

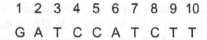

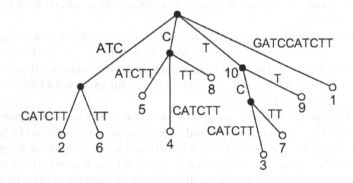

Fig. 4.2 A suffix tree for GATCCATCTT.

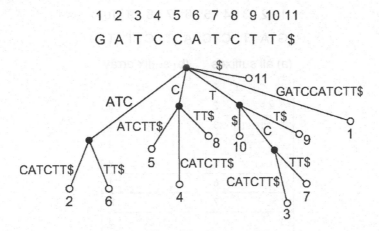

Fig. 4.3 A suffix tree for GATCCATCTT$.

and $B = b_1 b_2 \ldots b_n$, and a positive integer w. An exact word match of length w occurs in $a_i a_{i+1} \ldots a_{i+w-1}$ and $b_j b_{j+1} \ldots b_{j+w-1}$ if and only if the suffixes $a_i a_{i+1} \ldots a_m$ and $b_j b_{j+1} \ldots b_n$ share a common prefix of length at least w. With a suffix tree at hand, finding a common prefix becomes an easy job since all suffixes with a common prefix will share the path from the root that labels out that common prefix in the suffix tree. Not only does it work well for finding all exact word matches of a fixed length, it can also be used to detect all maximal word matches between two sequences or among several sequences by employing a so-called generalized suffix tree, which is a suffix tree for a set of sequences. Interested readers are referred to the book by Gusfield [85].

4.1.3 Suffix Arrays

A suffix array for sequence $A = a_1 a_2 \ldots a_m$ is an array of all suffixes of A in lexicographical order. Figure 4.4 constructs a suffix array for GATCCATCTT. At first glance, this conceptual representation seems to require quadratic space, but in fact the suffix array needs only linear space since it suffices to store only the starting positions for all sorted suffixes.

Recall that an exact word match of length w occurs in $a_i a_{i+1} \ldots a_{i+w-1}$ and $b_j b_{j+1} \ldots b_{j+w-1}$ if and only if the suffixes $a_i a_{i+1} \ldots a_m$ and $b_j b_{j+1} \ldots b_n$ share a common prefix of length at least w. Once a suffix array has been built, one can look up the table for any particular prefix by a binary search algorithm. This search can be done even more efficiently if some data structure for querying the longest common prefixes is employed.

1 2 3 4 5 6 7 8 9 10

G A T C C A T C T T

(a) all suffixes   (b) suffix array

GATCCATCTT	1	ATCCATCTT	2
ATCCATCTT	2	ATCTT	6
TCCATCTT	3	CATCTT	5
CCATCTT	4	CCATCTT	4
CATCTT	5	CTT	8
ATCTT	6	GATCCATCTT	1
TCTT	7	T	10
CTT	8	TCCATCTT	3
TT	9	TCTT	7
T	10	TT	9

Fig. 4.4 A suffix array for GATCCATCTT.

4.2 FASTA

FASTA uses a multi-step approach to finding local alignments. First, it finds runs of identities, and identifies regions with the highest density of identities. A parameter *ktup* is used to describe the minimum length of the identity runs. These runs of identities are grouped together according to their diagonals. For each diagonal, it locates the highest-scoring segment by adding up bonuses for matches and subtracting penalties for intervening mismatches. The ten best segments of all diagonals are selected for further consideration.

The next step is to re-score those selected segments using the scoring matrix such as PAM and BLOSUM, and eliminate segments that are unlikely to be part of the alignment. If there exist several segments with scores greater than the cutoff, they will be joined together to form a chain provided that the sum of the scores of the joined regions minus the gap penalties is greater than the threshold.

Finally, it considers the band of a couple of residues, say 32, centered on the chain found in the previous step. A banded Smith-Waterman method is used to deliver an optimal alignment between the query sequence and the database sequence.

Since FASTA was the first popular biological sequence database search program, its sequence format, called FASTA format, has been widely adopted. FASTA format is a text-based format for representing DNA, RNA, and protein sequences, where each sequence is preceded by its name and comments as shown below:

```
>HAHU Hemoglobin alpha chain - Human
VLSPADKTNVKAAWGKVGAHAGEYGAEALERMFLSFPTTK
TYFPHFDLSHGSAQVKGHGKKVADALTNAVAHVDDMPNAL
```

```
SALSDLHAHKLRVDPVNFKLLSHCLLVTLAAHLPAEFTPA
VHASLDKFLASVSTVLTSKYR
```

4.3 BLAST

The BLAST program is the most widely used tool for homology search in DNA and protein databases. It finds regions of local similarity between a query sequence and each database sequence. It also calculates the statistical significance of matches. It has been used by numerous biologists to reveal functional and evolutionary relationships between sequences and identify members of gene families.

The first version of BLAST was launched in 1990. It generates ungapped alignments and hence is called *Ungapped BLAST*. Seven years later, BLAST 2.0 came to the world. Two major products of BLAST 2.0 are Gapped BLAST and Position-Specific Iterated BLAST (PSI-BLAST). Gapped BLAST produces gapped alignments, yet it is able to run faster than the original one. PSI-BLAST can be used to find distant relatives of a protein based the profiles derived from the multiple alignments of the highest scoring database sequence segments with the query segment in iterative Gapped BLAST searches.

4.3.1 Ungapped BLAST

As discussed in Chapter 3, all possible pairs of residues are assigned their similarity scores when we compare biological sequences. For protein sequences, PAM or

	C	T	A	T	C	A	T	T	C	T	G
G	-4	-4	-4	-4	-4	-4	-4	-4	-4	-4	5
A	-4	-4	5	-4	-4	5	-4	-4	-4	-4	-4
T	-4	5	-4	5	-4	-4	5	5	-4	5	-4
C	5	-4	-4	-4	5	-4	-4	-4	5	-4	-4
C	5	-4	-4	-4	5	-4	-4	-4	5	-4	-4
A	-4	-4	5	-4	-4	5	-4	-4	-4	-4	-4
T	-4	5	-4	5	-4	-4	5	5	-4	5	-4
C	5	-4	-4	-4	5	-4	-4	-4	5	-4	-4
T	-4	5	-4	5	-4	-4	5	5	-4	5	-4
T	-4	5	-4	5	-4	-4	5	5	-4	5	-4

Fig. 4.5 A matrix of similarity scores for the pairs of residues of the two sequences GATCCATCTT and CTATCATTCTG.

BLOSUM substitution matrix is often employed, whereas for DNA sequences, an identity is given a positive score and a mismatch is penalized by a negative score. Figure 4.5 depicts the similarity scores of all the pairs of the residues of the two sequences GATCCATCTT and CTATCATTCTG, where an identity is given a score +5 and a mismatch is penalized by -4.

Let a sequence segment be a contiguous stretch of residues of a sequence. The score for the aligned segments $a_i a_{i+1} \ldots a_{i+\ell-1}$ and $b_j b_{j+1} \ldots b_{j+\ell-1}$ of length ℓ is the sum of the similarity scores for each pair of aligned residues (a_{i+k}, b_{j+k}) where $0 \leq k < \ell$. A maximal-scoring segment pair (MSP) is the highest scoring pair of segments of the same length chosen from the two sequences. Since its score is the highest, any stretch of this aligned segment pair will not increase the similarity score. In order to compute the MSP score, a straightforward approach is to compute the maximum-sum segment for each diagonal of the similarity scores matrix of the two sequences. Fix a diagonal, the maximum-sum segment can be found by a linear-time algorithm for the maximum-sum segment problem given in Section 2.4.2. Figure 4.6 locates a maximal-scoring segment pair in Figure 4.5.

However, there are $O(m+n)$ diagonals to be processed. If we apply the linear-time algorithm to all the diagonals, the resulting method takes the time proportional to the product of the lengths of the sequences. To speed up the computation, BLAST computes approximate MSPs, often referred to as high-scoring segment pairs (*HSPs*), in two phases. The first phase is to scan the database for hits, which are word pairs of length w with score at least T. The second phase is to extend each hit to see if it is contained within a segment pair whose score is no less than S.

Let us now explain these two phases in greater detail. In the first phase, BLAST seeks only segment pairs containing a *hit*, which is a word pair of length w with score at least T. For DNA sequences, these word pairs are exact word matches of fixed

		1 C	2 T	3 A	4 T	5 C	6 A	7 T	8 T	9 C	10 T	11 G
1	G	-4	-4	-4	-4	-4	-4	-4	-4	-4	-4	5
2	A	-4	-4	5	-4	-4	5	-4	-4	-4	-4	-4
3	T	-4	5	-4	5	-4	-4	5	5	-4	5	-4
4	C	5	-4	-4	-4	5	-4	-4	-4	5	-4	-4
5	C	5	-4	-4	-4	5	-4	-4	-4	5	-4	-4
6	A	-4	-4	5	-4	-4	5	-4	-4	-4	-4	-4
7	T	-4	5	-4	5	-4	-4	5	5	-4	5	-4
8	C	5	-4	-4	-4	5	-4	-4	-4	5	-4	-4
9	T	-4	5	-4	5	-4	-4	5	5	-4	5	-4
10	T	-4	5	-4	5	-4	-4	5	5	-4	5	-4

Fig. 4.6 A maximum-scoring segment pair of the two sequences GATCCATCTT and CTATCATTCTG.

length w, whereas for protein sequences, these word pairs are those fixed-length segment pairs who have a score no less than the threshold T.

Section 4.1 gives three methods for finding exact word matches between two sequences. Figure 4.7 depicts all the exact word matches of length three between the two sequences GATCCATCTT and CTATCATTCTG.

For protein sequences, we are not looking for exact word matches. Instead, a hit is a fixed-length segment pair having a score no less than the threshold T. A query word may be represented by several different words whose similarity scores with the query word are at least T.

The second phase of BLAST is to extend a hit in both directions (*diagonally*) to find a locally maximal-scoring segment pair containing that hit. It continues the extension in one direction until the score has dropped more than X below the maximum score found so far for shorter extensions. If the resulting segment pair has score at least S, then it is reported.

It should be noted that both the Smith-Waterman algorithm and BLAST asymptotically take the time proportional to the product of the lengths of the sequences. The speedup of BLAST comes from the reduced sample space size. For two sequences of lengths m and n, the Smith-Waterman algorithm involves $(n+1) \times (m+1)$ entries in the dynamic-programming matrix, whereas BLAST takes into account only those w-mers, whose number is roughly $mn/4^w$ for DNA sequences or $mn/20^w$ for protein sequences.

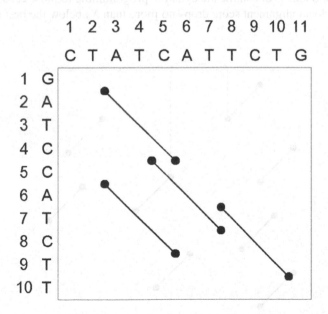

Fig. 4.7 Exact word matches of length three between the two sequences GATCCATCTT and CTATCATTCTG.

4.3.2 Gapped BLAST

Gapped BLAST uses a new criterion for triggering hit extensions and generates gapped alignment for segment pairs with "high scores."

It was observed that the hit extension step of the original BLAST consumes most of the processing time, say 90%. It was also observed that an HSP of interest is much longer than the word size w, and it is very likely to have multiple hits within a relatively short distance of one another on the same diagonal. Specifically, the two-hit method is to invoke an extension only when two non-overlapping hits occur within distance D of each other on the same diagonal (see Figure 4.8). These adjacent non-overlapping hits can be detected if we maintain, for each diagonal, the coordinate of the most recent hit found.

Another desirable feature of Gapped BLAST is that it generates gapped alignments explicitly for some cases. The original BLAST delivers only ungapped alignments. Gapped alignments are implicitly taken care of by calculating a joint statistical assessment of several distinct HSPs in the same database sequence.

A gapped extension is in general much slower than an ungapped extension. Two ideas are used to handle gapped extensions more efficiently. The first idea is to trigger a gapped extension only for those HSPs with scores exceeding a threshold S_g. The parameter S_g is chosen in a way that no more than one gap extension is invoked per 50 database sequences.

The second idea is to confine the dynamic programming to those cells for which the optimal local alignment score drops no more than X_g below the best alignment

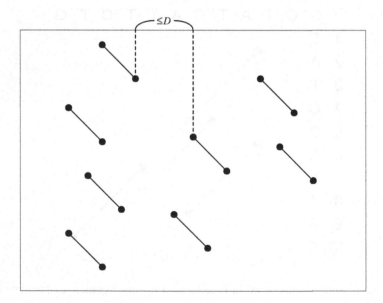

Fig. 4.8 Two non-overlapping hits within distance D of each other on the same diagonal.

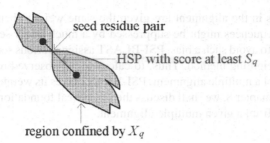

region confined by X_q

Fig. 4.9 A scenario of the gap extensions in the dynamic-programming matrix confined by the parameter X_g.

score found so far. The gapped extension for a selected HSP starts from a seed residue pair, which is a central residue pair of the highest-scoring length-11 segment pair along the HSP. If the HSP is shorter than 11, its central residue pair is chosen as the seed. Then the gapped extension proceeds both forward and backward through the dynamic-programming matrix confined by the parameter X_g (see Figure 4.9).

4.3.3 PSI-BLAST

PSI-BLAST runs BLAST iteratively with an updated scoring matrix generated automatically. In each iteration, PSI-BLAST constructs a position specific score matrix (PSSM) of dimension $\ell \times 20$ from a multiple alignment of the highest-scoring segments with the query segment of length ℓ. The constructed PSSM is then used to score the segment pairs for the next iteration. It has been shown that this iterative approach is often more sensitive to weak but biologically relevant sequence similarities.

PSI-BLAST collects, from the BLAST output, all HSPs with E-value below a threshold, say 0.01, and uses the query sequence as a template to construct a multiple alignment. For those selected HSPs, all database sequence segments that are identical to the query segment are discarded, and only one copy is kept for those database sequence segments with at least 98% identities. In fact, users can specify the maximum number of database sequence segments to be included in the multiple alignment. In case the number of HSPs with E-value below a threshold exceeds the maximum number, only those top ones are reported. A sample multiple alignment is given below:

```
query:    GVDIIIMGSHGKTNLKEILLGSVTENVIKKSNKPVLVVK
seq1:     GADVVVIGSR-NPSISTHLLGSNASSVIRHANLPVLVVR
seq2:     PAHMIIIASH-RPDITTYLLGSNAAAVVRHAECSVLVVR
seq3:     QAGIVVLGTVGRTGISAAFLGNTAEQVIDHLRCDLLVIK
```

If all segments in the alignment are given the same weight, then a small set of more divergent sequences might be suppressed by a much larger set of closely related sequences. To avoid such a bias, PSI-BLAST assigns various sequence weights by a sequence weighting method. Thus, to calculate the observed residue frequencies of a column of a multiple alignment, PSI-BLAST takes its weighted frequencies into account. In Chapter 8, we shall discuss the theoretical foundation for generating scoring matrices from a given multiple alignment.

4.4 BLAT

BLAT is short for "BLAST-like alignment tool." It is often used to search for database sequences that are closely related to the query sequences. For DNA sequences, it aims to find those sequences of length 25 bp or more and with at least 95% similarity. For protein sequences, it finds those sequences of length 20 residues or more and with at least 80% similarity.

A desirable feature is that BLAT builds an index of the whole database and keeps it in memory. The index consists of non-overlapping K-mers and their positions in the database. It excludes those K-mers that are heavily involved in repeats. DNA BLAT sets K to 11, and protein BLAT sets K to 4. The index requires a few gigabytes of RAM and is affordable for many users. This feature lets BLAT scan linearly through the query sequence, rather than scan linearly through the database.

BLAT builds a list of hits by looking up each overlapping K-mer of the query sequence in the index (see Figure 4.10). The hits can be single perfect word matches or near perfect word matches. The near perfect mismatch option allows one mismatch in a hit. Given a K-mer, there are $K \times (|\Sigma| - 1)$ other possible K-mers that match it in all but one position, where $|\Sigma|$ is the alphabet size. Therefore, the near perfect mismatch option would require $K \times (|\Sigma| - 1) + 1$ lookups per K-mer of the query sequences.

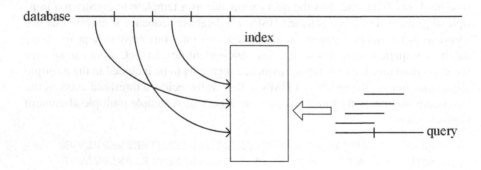

Fig. 4.10 The index consists of non-overlapping K-mers in the database, and each overlapping K-mer of the query sequence is looked up for hits.

BLAT identifies homologous regions of the database by clumping hits as follows. The hits are distributed into buckets of 64k according to their database positions. In each bucket, hits are bundled into a clump if they are within the gap limit on the diagonal and the window limit on the database coordinate. To smooth the boundary cut, the hits and clumps within the window limit of the next bucket are passed on for possible clumping and extension. If a clump contains a certain amount of hits, then it defines a region of the database that are homologous to the query sequence. If two homologous regions of the database are within 300 nucleotides or 100 amino acids, they are merged as one. Finally, each homologous region is flanked with a few hundred additional residues on both sides.

The alignment stage of BLAT delivers alignments in the homologous regions found in the clumping process. The alignment procedures for nucleotide sequences and protein sequences are different. Both of them work well for aligning sequences that are closely related.

For nucleotide alignment, the alignment procedure works as follows. It stretches each K-mer as far as possible allowing no mismatches. An extended K-mer forms a hit if it is unique or exceeds a certain length. Overlapping hits are merged together. To bridge the gaps among hits, a recursive procedure is employed. A recursion continues if it finds no additional hits using a smaller K or the gap is no more than five nucleotides.

For protein alignment, the alignment procedure works as follows. The hits are extended into high-scoring segment pairs (HSPs) by giving a bonus score $+2$ to a match and penalizing a mismatch by -1. Let H_1 and H_2 be two HSPs that are non-overlapping in both the database and the query sequences, and assume that H_1 precedes H_2. An edge is added to connect from H_1 to H_2, where the edge weight is the score of H_2 minus the gap cost based on the distance between H_1 and H_2. For those overlapping HSPs, a cutting point is chosen to maximize the score of the joint HSP. Then a dynamic-programming method is used to find the maximal-scoring path, *i.e.*, the maximal-scoring alignment, in the graph one by one until all HSPs are reported in some alignment.

4.5 PatternHunter

As discussed in Section 4.3, BLAST computes HSPs by extending so-called "hits" or "seeds" between the query sequence and the database sequence. The seeds used by BLAST are short contiguous word matches. Some homologous regions might be missed if they do not contain any seed.

An advanced homology search program named PatternHunter has been developed to enhance the sensitivity by finding short word matches under a spaced seed model. A spaced seed is represented as a binary string of *length l*, where a "1" bit at a position means that a base match is required at the position, and a "$*$" bit at a position means that either a base match or mismatch is acceptable at the position. The number of 1 bits in a spaced seed is the *weight* of the seed. For example, the words ACGTC and ATGAC form a word match under spaced seed 1*1*1, but not under 11**1. Note that BLAST simply uses a consecutive model that consists of consecutive 1s, such as 11111.

In general, a spaced seed model shares fewer 1s with any of its shifted copies than the contiguous one. Define the number of overlapping 1s between a model and its shifted copy as the number of 1s in the shifted copy that correspond to 1s in the model. The number of non-overlapping 1s between a model and its shifted copy is the weight of the model minus the number of overlapping 1s. If there are more non-overlapping 1s between the model and its shifted copy, then the conditional probability of having another hit given one hit is smaller, resulting in higher sensitivity. For rigorous analysis of the hit probabilities of spaced seeds, the reader is referred to Chapter 6.

A model of length l has $l-1$ shifted copies. For a model π of length l, the sum of overlapping hit probabilities between the model and each of its shifted copies, $\phi(\pi, p)$, is calculated by the equation

$$\phi(\pi, p) = \sum_{i=1}^{l-1} p^{n_i}, \tag{4.1}$$

where p is the similarity level and n_i denotes the number of non-overlapping 1s between the model π and its ith shifted copy. Figure 4.11 computes $\phi(\pi, p)$ for $\pi=1*11*1$.

Both empirical and analytical studies suggest that, among all the models of fixed length and weight, a model is more sensitive if it has a smaller sum of overlapping hit probabilities. A model of length l and weight w is an *optimal* model if its ϕ value is minimum among all models of length l and weight w. For example, the spaced seed model 111*1**1*1**11*111 used in PatternHunter is an optimal one for length 18 and weight 11 with similarity level $p = 0.7$.

In order to calculate the value of ϕ for a spaced seed model, we need to count the number of non-overlapping 1s between the model and each of its shifted copies, which can be computed by subtracting the number of overlapping 1s from the weight w. This can be done by a straightforward dynamic-programming method, which takes $O(\ell^2)$ time to compute ϕ for any model of length ℓ. By observing that at most $O(\ell)$ bit overlaps differ for two models with only one pair of * and 1 swapped, one

i	i^{th} shifted copy	non-overlapping 1s	overlapping hit probability
0	1*11*1		
1	1*11*1	3	p^3
2	1*11*1	2	p^2
3	1*11*1	2	p^2
4	1*11*1	4	p^4
5	1*11*1	3	p^3

Fig. 4.11 Calculation of the sum of overlapping hit probabilities between the model π and each of its shifted copies, $\phi(\pi, p) = p^3 + p^2 + p^2 + p^4 + p^3$, for $\pi=1*11*1$.

can develop an $O(\ell)$ time algorithm for updating the numbers of non-overlapping 1s for a swapped model. This suggests a practical computation method for evaluating ϕ values of all models by orderly swapping one pair of $*$ and 1 at a time.

Given a spaced seed model, how can one find all the hits? Recall that only those 1s account for a match. One way is to employ a hash table like Figure 4.1 for exact word matches but use a spaced index instead of a contiguous index. Let the spaced seed model be of length ℓ and weight w. For DNA sequences, a hash table of size 4^w is initialized. Then we scan the sequence with a window of size ℓ from left to right. We extract the residues corresponding to those 1s as the index of the window (see Figure 4.12).

Once the hash table is built, the lookup can be done in a similar way. Indeed, one can scan the other sequence with a window of size ℓ from left to right and extract the residues corresponding to those 1s as the index for looking up the hash table for hits.

Hits are extended diagonally in both sides until the score drops by a certain amount. Those extended hits with scores exceeding a threshold are collected as

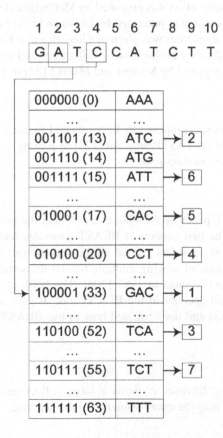

Fig. 4.12 A hash table for GATCCATCTT under a weight three model 11*1.

high-scoring segment pairs (HSPs). As for the gap extension of HSPs, a red-black tree with diagonals as the key is employed to manipulate the extension process efficiently.

4.6 Bibliographic Notes and Further Reading

Twenty years ago, it would have been a legend to find similarities between two sequences. However, nowadays it would be a great surprise if we cannot find similarities between a newly sequenced biomolecular sequence and the GenBank database. FASTA and BLAST were the ones that boosted this historical change. In Table C.2 of Appendix C, we compile a list of homology search tools.

4.1

The first linear-time suffix tree was given by Weiner [202] in 1973. A space-efficient linear-time construction was proposed by McCreight [135] in 1976. An on-line linear-time construction was presented by Ukkonen [191] in 1995. A review of these three linear-time algorithms was given by Giegerich and Kurtz [75]. Gusfield's book [85] gave a very detailed explanation of suffix trees and their applications.

Suffix arrays were proposed by Manber and Myers [134] in 1991.

4.2

The FASTA program was proposed by Pearson and Lipman [161] in 1988. It improved the sensitivity of the FASTP program [128] by joining several initial regions if their scores pass a certain threshold.

4.3

The original BLAST paper by Altschul et al. [7] was the most cited paper published in the 1990s. The first version of BLAST generates ungapped alignments, whereas BLAST 2.0 [8] considers gapped alignments as well as position-specific scoring schemes. The idea of seeking multiple hits on the same diagonal was first proposed by Wilbur and Lipman in 1983 [204]. The book of Korf, Yandell, and Bedell [116] gave a full account of the BLAST tool. It gave a clear understanding of BLAST programs and demonstrated how to use BLAST for taking the full advantage.

4.4

The UCSC Genome Browser contains a large collection of genomes [104]. BLAT [109] is used to map the query sequence to the genome.

4.5

Several variants of PatternHunter [131] are also available to the public. For instance, PatternHunter II [123] improved the sensitivity even further by using multiple spaced seeds, and tPatternHunter [112] was designed for doing protein-protein, translated protein-DNA, and translated DNA-DNA homology searches. A new version of BLASTZ also adapted the idea of spaced models [177] by allowing one transition (A–G, G–A, C–T or T–C) in any position of a seed.

Some related but different *spaced* approaches have been considered by others. Califano and Rigoutsos [37] introduced a new index generating mechanism where k-tuples are formed by concatenating non-contiguous subsequences that are spread over a large portion of the sequence of interest. The first stage of the multiple filtration approach proposed by Pevzner and Waterman [167] uses a new technique to preselect similar m-tuples that allow a few number of mismatches. Buhler [33] uses locality-sensitive hashing [96] to find K-mers that differ by no more than a certain number of substitutions.

Chapter 5
Multiple Sequence Alignment

Aligning simultaneously several sequences is among the most important problems in computational molecular biology. It finds many applications in computational genomics and molecular evolution.

This chapter is divided into five sections. Section 5.1 gives a definition of multiple sequence alignment and depicts a multiple alignment of three sequences.

There are several models for assessing the score of a given multiple sequence alignment. The most popular ones are sum-of-pairs (SP), tree alignment, and consensus alignment. In Section 5.2, we focus our discussion on the SP alignment scoring scheme.

Section 5.3 considers the problem of aligning three sequences based on the SP alignment model. An exact multiple alignment algorithm is given for such a problem.

For many applications of multiple alignment, more efficient heuristic methods are often required. Among them, most methods adopt the approach of "progressive" pairwise alignments introduced in Section 5.4. It iteratively merges the most similar pairs of sequences/alignments following the principle "once a gap, always a gap."

Finally, we conclude the chapter with the bibliographic notes in Section 5.5.

5.1 Aligning Multiple Sequences

Simultaneous alignment of several sequences is among the most important problems in computational molecular biology. Its purpose is to reveal the biological relationship among multiple sequences. For example, it can be used to locate conservative regions, study gene regulation, and to infer evolutionary relationship of genes or proteins.

Recall the definition of a pairwise alignment given in Chapter 1.2.1. An *alignment* of two sequences is obtained by inserting some number (perhaps 0) of spaces, denoted by dashes, in each sequence to yield padded sequences of equal length, then placing the first padded sequence above the other. To emphasize that all sequence

S_1:   TTATTTCACC-----CTTATATCA
S_2:   TCCTTTCA--------TGATATCA
S_3:   T--TTTCACCGACATCAGATAAAA

Fig. 5.1 A sample of multiple sequence alignment.

entries are required to appear in the alignment, we use the term *global* (as opposed to *local*). Each column of an alignment is called an *aligned pair*. In general, we require that an alignment does not contain two spaces in a column, which we call the *null column*. In context where null columns are permitted the term *quasi-alignment* is used to emphasize that the ban on null columns has been temporarily lifted.

Assume that we are given $S_1, S_2, \ldots, S_m$, each of which is a sequence of "letters." A multiple alignment of these sequences is an $m \times n$ array of letters and dashes, such that no column consisting entirely of dashes, and removing dashes from row i leaves the sequence S_i for $1 \leq i \leq m$. For each pair of sequences, say S_i and S_j, rows i and j of the m-way alignment constitute a pairwise quasi-alignment of S_i and S_j; removing any null columns produces a pairwise alignment of these sequences. Figure 5.1 gives a multiple alignment of three sequences:

5.2 Scoring Multiple Sequence Alignment

For any two given sequences, there are numerous alignments of those sequences. To make explicit the criteria for preferring one alignment over another, we define a score for each alignment. The higher the score is, the better the alignment is. Let us review the scoring scheme given in Section 1.3. First, we assign a score denoted $\sigma(x,y)$ to each aligned pair $\begin{pmatrix} x \\ y \end{pmatrix}$. In the cases that x or y is a space, $\sigma(x,y) = -\beta$. Score function σ depends only on the contents of the two locations, not their positions within the sequences. Thus, $\sigma(x,y)$ does not depend on where the particular symbols occur. However, it should be noted that there are situations where position-dependent scores are quite appropriate. Similar remarks hold for the gap penalties defined below.

The other ingredient for scoring pairwise alignments is a constant *gap-opening penalty*, denoted α, that is assessed for each gap in the alignment; a *gap* is defined as a run of spaces in a row of the alignment that is terminated by either a non-space symbol or an end of the row. Gap penalties are charged so that a single gap of length, say, k will be preferred to several gaps of total length k, which is desirable since a gap can be created in a single evolutionary event. Occasionally, a different scoring criterion will be applied to *end-gaps*, i.e., gaps that are terminated by an end of the row. The score of an alignment is defined as the sum of σ values for all aligned pairs, minus α times the number of gaps.

Selection of the scoring parameters σ and α is often a major factor affecting the usefulness of the computed alignments. Ideally, alignments are determined in such a way that sequence regions serving no important function, and hence evolving freely, should not align, whereas regions subject to purifying selection retain sufficient similarity that they satisfy the criteria for alignment. The chosen alignment scoring scheme determines which regions will be considered non-aligning and what relationships will be assigned between aligning regions. Appropriateness of scoring parameters depends on several factors, including evolutionary distance between the species being compared.

When simultaneously aligning more than two sequences, we want knowledge of appropriate parameters for pairwise alignment to lead immediately to appropriate settings for the multiple-alignment scoring parameters. Thus, one might desire a scoring scheme for multiple alignments that is intimately related to their induced pairwise alignment scores. Of course, it is also necessary that the approach be amenable to a multiple-alignment algorithm that is reasonably efficient with computer resources, *i.e.*, time and space.

There are several models for assessing the score of a given multiple sequence alignment. The most popular ones are sum-of-pairs (SP), tree alignment, and consensus alignment. We focus our discussion on the SP alignment scoring scheme.

To attain this tight coupling of pairwise and multiple alignment scores at a reasonable expense, many multiple alignment tools have adopted the *SP* substitution scores and quasi-natural gap costs, as described by Altschul [2]. Some notation will help for a precise description of these ideas.

Scores for multiple alignments are based on pairwise alignment scores, which we described above. With an m-way alignment Π, we would like to determine appropriate parameters for the score, say $Score_{i,j}$, for pairwise alignments between S_i and S_j (*i.e.*, the ith and jth sequences), then set

$$(SP)\ Score(\Pi) = \sum_{i<j} Score_{i,j}(\Pi_{i,j}),$$

where $\Pi_{i,j}$ is the pairwise alignment of S_i and S_j induced by Π (see Figure 5.2).

$\Pi_{1,2}$
```
S1:    TTATTTCACCCTTATATCA
S2:    TCCTTTCA---TGATATCA
```

$\Pi_{1,3}$
```
S1:    TTATTTCACC-----CTTATATCA
S3:    T--TTTCACCGACATCAGATAAAA
```

$\Pi_{2,3}$
```
S2:    TCCTTTCA--------TGATATCA
S3:    T--TTTCACCGACATCAGATAAAA
```

Fig. 5.2 Three pairwise alignments induced by the multiple alignment in Figure 5.1.

S_1:   TTATTTCACC-----CTTATATCA
S_2:   TCCTTTCA--------TGATATCA

Fig. 5.3 A quasi-alignment of Π (in Figure 5.1) projected on S_1 and S_2 without discarding null columns.

The projected substitution costs of SP-alignments can be computed easily. However, as noted in Altschul [2], strictly computing the imposed affine gap costs results in undesirable algorithmic complexity. The complications come from the fact that we may have to save a huge number of the relevant histories in order to decide if we need to charge a gap opening-up penalty for a given deletion (or insertion) pair. Altschul further observed that this complexity of saving all possible relevant histories can be reduced dramatically if for every pair of rows of the m-way alignment we assess an additional gap penalty for each "quasi-gap," defined as follows. Fix a pair of rows and consider a gap, G, in the corresponding pairwise quasi-alignment, i.e., a run of consecutive gap symbols occurring in one of the rows (the run should be extended in both directions until it hits a letter or the end of the sequence). If at least one space in G is aligned with a letter in the other row, then G corresponds to a gap in the pairwise alignment (i.e., after discarding null columns), and hence is penalized. The other possibility is that every space in G is aligned with a space in the other sequence. If the gap in the other sequence starts strictly before and ends strictly after G, then G is called a *quasi-gap* and is penalized. For example, the gap in S_2 of Figure 5.3 is a quasi-gap in a projected alignment without discarding null columns. In $\Pi_{1,2}$ of Figure 5.2, there is only one deletion gap counted. But in practical implementation, we might assess two gap penalties since an additional quasi-gap penalty might be imposed. If either end of G is aligned to an end of the gap in the other sequence, then the gap is not penalized.

In summary, a multiple alignment is scored as follows. For each pair of rows, say rows i and j, fix appropriate substitution scores $\sigma_{i,j}$ and a gap cost $\alpha_{i,j}$. Then the score for the multiple alignment is determined by equation (SP), where each $Score_{i,j}(\Pi_{i,j})$ is found by adding the σ values for non-null columns of the pairwise quasi-alignment, and subtracting a gap penalty α for each gap and each quasi-gap.

5.3 An Exact Method for Aligning Three Sequences

The pairwise alignment algorithms introduced in Chapter 3 can be easily extended to align for more than two sequences. Consider the problem of aligning three sequences $A = a_1 a_2 \ldots a_{n_1}$, $B = b_1 b_2 \ldots b_{n_2}$, and $C = c_1 c_2 \ldots c_{n_3}$ based on the SP alignment model. Let x, y and z be any alphabet symbol or a gap symbol. Assume that a simple scoring scheme for pairwise alignment is imposed where a score $\chi(x,y)$ is defined for each aligned pair $\begin{pmatrix} x \\ y \end{pmatrix}$. Let $\phi(x,y,z)$ be the score of an aligned column $\begin{pmatrix} x \\ y \\ z \end{pmatrix}$. The score $\phi(x,y,z)$ can be computed as the sum of $\chi(x,y)$, $\chi(x,z)$, and $\chi(y,z)$.

Let $S[i,j,k]$ denote the score of an optimal alignment of $a_1 a_2 \ldots a_i$, $b_1 b_2 \ldots b_j$, and $c_1 c_2 \ldots c_k$. With proper initializations, $S[i,j,k]$ for $1 \leq i \leq n_1$, $1 \leq j \leq n_2$, and $1 \leq k \leq n_3$ can be computed by the following recurrence.

$$S[i,j,k] = \max \begin{cases} S[i-1,j,k] + \phi(a_i,-,-), \\ S[i,j-1,k] + \phi(-,b_j,-), \\ S[i,j,k-1] + \phi(-,-,c_k), \\ S[i,j-1,k-1] + \phi(-,b_j,c_k), \\ S[i-1,j,k-1] + \phi(a_i,-,c_k), \\ S[i-1,j-1,k] + \phi(a_i,b_j,-), \\ S[i-1,j-1,k-1] + \phi(a_i,b_j,c_k). \end{cases}$$

The value $S[n_1, n_2, n_3]$ is the score of an optimal multiple alignment of A, B, and C. The three-dimensional dynamic-programming matrix contains $O(n_1 n_2 n_3)$ entries, and each entry takes the maximum value from the $2^3 - 1 = 7$ possible entering edges. All possible combinations of ϕ values can be computed in advance. Thus, we can align three sequences of lengths n_1, n_2 and n_3 in $O(n_1 n_2 n_3)$ time.

Following this approach, one can easily derive an $O(n^m 2^m)$-time algorithm for constructing m-way alignment of length n. This exact method in general requires too much time and space to be practical for DNA sequences of average length. Not to mention that there are a lot more possible entering edges (configurations) for each entry if affine gap penalties or affine quasi-gap penalties are used. Furthermore, the multiple sequence alignment has been shown to be NP-hard by Wang and Jiang [195], meaning that there is no polynomial-time algorithm for it unless NP=P.

Despite the intractability of the multiple alignment problem, some researchers proposed "efficient" exact methods by pruning the dynamic-programming matrix with some optimal score lower bound. These exact methods have been proved to be useful in certain context.

5.4 Progressive Alignment

For many applications of multiple alignment, more efficient heuristic methods are often required. Among them, most methods adopt the approach of "progressive" pairwise alignments proposed by Feng and Doolittle [69]. It iteratively merges the most similar pairs of sequences/alignments following the principle "once a gap, always a gap." Thus, later steps of the process align two "sequences," one or both of which can themselves be an alignment, *i.e.*, sequence of fixed-height columns.

In aligning two pairwise alignments, the columns of each given pairwise alignment are treated as "symbols," and these sequences of symbols are aligned by

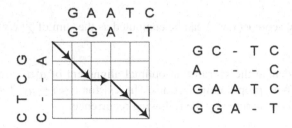

Fig. 5.4 Aligning two alignments.

padding each sequence with appropriate-sized columns containing only dash symbols. It is quite helpful to recast the problem of aligning two alignments as an equivalent problem of finding a maximum-scoring path in an alignment graph. For example, the path depicted in Figure 5.4 corresponds to a 4-way alignment. This alternative formulation allows the problem to be visualized in a way that permits the use of geometric intuition. We find this visual imagery critical for keeping track of the low-level details that arise in development and implementation of alignment algorithms.

Each step of the progressive alignment procedure produces an alignment that is highest-scoring relative to the chosen scoring scheme subject to the constraint that columns of the two smaller alignments being combined are treated as indivisible "symbols." Thus, the relationships between entries of two of the original sequences are fixed at the first step that aligns those sequences or alignments containing those sequences.

For that reason, it is wise to first compute the pairwise alignments that warrant the most confidence, then combine those into multiple alignments. Though each step is performed optimally, there is no guarantee that the resulting multiple alignment is highest-scoring over all possible ways of aligning the given sequences. An appropriate order for progressive alignment is very critical for the success of a multiple alignment program. This order can either be determined by the guide tree constructed from the distance matrix of all pairs of sequences, or can be inferred directly from an evolutionary tree for those sequences. In any case, the progressive alignment algorithm invokes the "generalized" pairwise alignment $m - 1$ times for constructing an m-way alignment, and its time complexity is roughly the order of the time for computing all $O(m^2)$ pairwise alignments.

5.5 Bibliographic Notes and Further Reading

In spite of the plethora of existing ideas and methods for multiple sequence alignment, it remains as an important and exciting line of investigation in computational molecular biology. Recently, Miller et al. [140] compiled a set of alignments of 28 vertebrate genome sequences in the UCSC Genome Browser [104].

5.1

We compile a list of multiple alignment tools in Table C.3 of Appendix C. There are a few multiple alignment benchmarks available for testing such as BAl-iBASE [20], PREFAB [62], and SMART [122]. Thompson et al. [190] gave the first systematic comparison of multiple alignment programs using BAliBASE benchmark dataset.

5.2

A recent survey by Notredame [154] divides the scoring schemes in two categories. Matrix-based methods, such as ClustalW [120, 189], Kalign [121], and MUSCLE [62], use a substitution matrix to score matches. On the other hand, consistency-based methods, such as T-Coffee [155], MUMMALS [163], and Prob-Cons [59], compile a collection of pairwise alignments and produce a position-specific substitution matrix to judge the consistency of a given aligned pair. An alternative is to score a multiple alignment by using a consensus sequence that is derived from the consensus of each column of the alignment.

Besides these two categories, there are some other scoring schemes. For example, DIALIGN [143] focuses on scoring complete segments of sequences.

5.3

A straightforward multiple sequence alignment algorithm runs in exponential time. More "efficient" exact methods can be found in [38, 84, 127]. In fact, it has been shown to be NP-hard by Wang and Jiang [195]. Bafna et al. [18] gave an algorithm with approximation ratio $2 - \ell/m$ for any fixed ℓ.

5.4

A heuristic progressive alignment approach was proposed by Feng and Doolittle [69]. It iteratively merges the most similar pairs of sequences/alignments following the principle "once a gap, always a gap."

Surprisingly, the problem of aligning two SP-alignments under affine gap penalties has been proved to be NP-hard [107, 132]. However, it becomes tractable if affine quasi-gap penalties are used [2].

ClustalW [120, 189] is the most popular multiple alignment program. It works in three stages. In the first stage, all pairs of sequences are aligned, and a distance matrix is built. The second stage constructs a guide tree from the distance matrix. Finally, in the third stage, the sequences as well as the alignments are aligned progressively according to the order in the guide tree.

YAMA [41] is a multiple alignment program for aligning long DNA sequences. At each progressive step, it implements the generalized Hirschberg linear-space algorithm and maximizes the sum of pairwise scores with affine quasi-gap penalties. To increase efficiency, a step of the progressive alignment algorithm can be constrained to the portion of the dynamic-programming grid lying between two boundary lines. Another option is to consider constrained alignments consisting of aligned pairs in nearly optimal alignments. A set of "patterns" is specified (perhaps the consensus sequences for transcription factor binding sites); YAMA selects, from among all alignments with the highest score, an alignment with the largest number of conserved blocks that match a pattern.

MUSCLE [62, 63] is a very efficient and accurate multiple alignment tool. It uses a strategy similar to PRRP [80] and MAFFT [106]. It works in three stages. Stage 1 is similar to ClustalW. A guide tree is built based on the distance matrix, which is derived from pairwise similarities. A draft progressive alignment is built according the constructed guide tree. In Stage 2, MUSCLE computes a Kimura distance matrix [111] using the pairwise identity percentage induced from the multiple alignment. It then builds a new guide tree based on such a matrix, compares it with the previous tree, and re-aligns the affected subtrees. This stage may be iterated if desired. Stage 3 iteratively refines the multiple alignment by re-aligning the profiles of two disjoint subsets of sequences derived from deleting an edge from the tree constructed in Stage 2. This refinement is a variant of the method by Hirosawa et al. [90]. If the new alignment has a better score, save it. The refinement process terminates if all edges incur no changes or a user-specified maximum number of iterations has been reached.

PART II. THEORY

Chapter 6
Anatomy of Spaced Seeds

BLAST was developed by Lipman and his collaborators to meet demanding needs of homology search in the late 1980s. It is based on the filtration technique and is multiple times faster than the Smith-Waterman algorithm. It first identifies short exact matches (called seed matches) of a fixed length (usually 11 bases) and then extends each match to both sides until a drop-off score is reached. Motivated by the success of BLAST, several other seeding strategies were proposed at about the same time in the early 2000s. In particular, PatternHunter demonstrates that an optimized spaced seed improves sensitivity substantially. Accordingly, elucidating the mechanism that confers power to spaced seeds and identifying good spaced seeds are two new issues of homology search.

This chapter is divided into six sections. In Section 6.1, we define spaced seeds and discuss the trade-off between sensitivity and specificity for homology search.

The sensitivity and specificity of a seeding-based program are largely related to the probability that a seed match is expected to occur by chance, called the hit probability. Here we study analytically spaced seeds in the Bernoulli sequence model defined in Section 1.6. Section 6.2 gives a recurrence relation system for calculating hit probability.

In Section 6.3, we investigate the expected distance μ between adjacent non-overlapping seed matches. By estimating μ, we further discuss why spaced seeds are often more sensitive than the consecutive seed used in BLAST.

A spaced seed has a larger span than the consecutive seed of the same weight. As a result, it has less hit probability in a small region but surpass the consecutive seed for large regions. Section 6.4 studies the hit probability of spaced seeds in asymptotic limit. Section 6.5 describes different methods for identifying good spaced seeds.

Section 6.6 introduces briefly three generalizations of spaced seeds: transition seeds, multiple spaced seeds, and vector seeds.

Finally, we conclude the chapter with the bibliographic notes in Section 6.7.

6.1 Filtration Technique in Homology Search

Filtration is a powerful strategy for homology search as exemplified by BLAST. It identifies short perfect matches defined by a fixed pattern between the query and target sequences and then extends each match to both sides for local alignments; the obtained local alignments are scored for acceptance.

6.1.1 Spaced Seed

The pattern used in filtration strategy is usually specified by one or more strings over the alphabet $\{1, *\}$. Each such string is a *spaced seed* in which 1s denote matching positions. For instance, if the seed $P = 11 * 1 * *11$ is used, then the segments examined for match in the first stage span 8 positions and the program only checks whether they match in the 1st, 2nd, 4th, 7th, and 8th positions or not. If the segments match in these positions, they form a perfect match. As we have seen in this example, the positions specified by *s are irrelevant and sometimes called *don't care* positions.

The most natural seeds are those in which all the positions are matching positions. These seeds are called the *consecutive seeds*. They are used in BLASTN. WABA, another homology search program, uses spaced seed $11 * 11 * 11 * 11 * 11$ to align gene-coding regions. The rationale behind this seed is that mutation in the third position of a codon usually does not affect the function of the encoded amino acid and hence substitution in the third base of a codon is irrelevant. PatternHunter uses a rather unusual seed $111 * 1 * *1 * 1 * *11 * 111$ as its default seed. As we shall see later, this spaced seed is much more sensitive than the BLASTN's default seed although they have the same number of matching positions.

6.1.2 Sensitivity and Specificity

The efficiency of a homology search program is measured by sensitivity and specificity. Sensitivity is the true positive rate, the chance that an alignment of interest is found by the program; specificity is one minus the false positive rate; the false positive rate is the chance that an alignment without any biological meaning is found in comparison of random sequence.

Obviously, there is a trade-off between sensitivity and specificity. In a program powered with a spaced seed in the filtration phase, if the number of the matching positions of the seed is large, the program is less sensitive, but more specified. On the other hand, lowering the number of the matching positions increases sensitivity, but decreases the specificity. What is not obvious is that the positional structure of a spaced seed affects greatly the sensitivity and specificity of the program.

To study analytically the sensitivity and specificity of a spaced seed, a model of ungapped alignments is needed. If we use 1s and 0s to represent matches and mismatches, an ungapped alignment is just a binary sequence. A frequently used ungapped alignment model is the kth-order Markov chain process $\mathcal{M}$. It is specified by giving the probability that 1 (or a match) is present at any position given the values of the k preceding bits. The 0th-order Markov chain model, also called the Bernoulli sequence model, describes the alignment's overall degree of conservation, and a higher-order model can refect specific patterns of conservation. For example, an alignment between two gene-coding sequences often exhibits a pattern of two matches followed by a mismatch because the point mutation occurring at the third base position of a codon is usually silent. Therefore, a 3rd-order Markov chain model is a better model for alignments between coding sequences. A kth-order Markov chain model can be trained on biological meaningful alignments obtained from real DNA sequences of interest.

Here, we study the sensitivity of the spaced seeds in the Bernoulli sequence model. Although this simple model ignores some biologically information, such as the frequencies of transitions and transversions, it suffices to serve our purpose. Most of the theorems proved in this chapter can generalize to the high-order Markov chain models in a straightforward manner.

Let π be a spaced seed. Its *length* or span is written $|\pi|$. The number of the matching positions of π is called its *weight* and written w_π. We assume that the first and last positions of π are matching positions because a seed that does not satisfy the condition is equivalent to a shorter one.

Assume π has matching positions $i_1, i_2, \cdots, i_{w_\pi}$ It hits an alignment A of interest if every position of A inspected at an offset j must contain match bases. That is, in the bit-representation R of A, $R[j - i_{w_\pi} + i_k + 1] = 1$ for all $1 \leq k \leq w_\pi$. Recall that a homology search program powered with seed π first enumerates all the pairs of positions in the aligned sequences that exhibit seed matches. The sensitivity and specificity of the program largely depend on the probability of π hitting a random binary sequence, which closely related to the average number of non-overlapping hits as we shall see later. Therefore, to design effective alignment program, one looks for the seed π that maximizes the hit probability on a random sequence.

6.2 Basic Formulas on Hit Probability

In the Bernoulli sequence model, an ungapped alignment is modeled as a random binary sequence in which 1 stands for match and is generated with a probability. In the rest of this chapter, p always denotes the probability that 1 is generated at a position of a random sequence and $q = 1 - p$.

Let R be an infinite random binary sequence whose letters are indexed from 0. We use $R[k]$ to denote the kth symbol of R and $R[i, j]$ the substring from position i to position j inclusively for $i < j$.

For a spaced seed π, all matching positions form the following ordered set:

$$\mathscr{RP}(\pi) = \{i_1 = 0, i_2, \cdots, i_{w_\pi} = |\pi| - 1\}. \tag{6.1}$$

The seed π is said to hit R at position j if and only if $R[j - |\pi| + i_k + 1] = 1$ for all $1 \leq k \leq w_\pi$. Here, we use the ending position as the hit position following the convention in the renewal theory.

Let A_i denote the event that π hits R at position i and $\bar{A}_i$ the complement of A_i. We use π_i to denote the probability that π first hits R at position $i - 1$, that is,

$$\pi_i = \Pr[\bar{A}_0 \bar{A}_1 \cdots \bar{A}_{i-2} A_{i-1}].$$

We call π_i the *first hit probability*. Let $\Pi_n := \Pi_n(p)$ denote the probability that π hits $R[0, n-1]$ and $\bar{\Pi}_n := 1 - \Pi_n$. We call Π_n the *hit probability* and $\bar{\Pi}_n$ the *non-hit probability* of π.

For each $0 \leq n < |\pi| - 1$, trivially, $A_n = \emptyset$ and $\Pi_n = 0$. Because the event $\cup_{0 \leq i \leq n-1} A_i$ is the disjoint union of the events

$$\cup_{0 \leq i \leq n-2} A_i$$

and

$$\overline{\cup_{0 \leq i \leq n-2} A_i} A_{n-1} = \bar{A}_0 \bar{A}_1 \cdots \bar{A}_{n-2} A_{n-1},$$

$$\Pi_n = \pi_1 + \pi_2 + \cdots + \pi_n,$$

or equivalently,

$$\bar{\Pi}_n = \pi_{n+1} + \pi_{n+2} + \cdots \tag{6.2}$$

for $n \geq |\pi|$.

Example 6.1. Let θ be the consecutive seed of weight w. If θ hits random sequence R at position $n - 1$, but not at position $n - 2$, then $R[n - w, n - 1] = 11 \cdots 1$ and $R[n - w - 1] = 0$. As a result, θ cannot hit R at positions $n - 3, n - 4, \ldots, n - w - 1$. This implies

$$\theta_n = \Pr[\bar{A}_0 \bar{A}_1 \cdots \bar{A}_{n-2} A_{n-1}] = p^w (1 - p) \bar{\Theta}_{n-w-1}, \quad n \geq w + 1.$$

By formula (6.2), its non-hit probability satisfies the following recurrence relation

$$\bar{\Theta}_n = \bar{\Theta}_{n-1} - p^w (1 - p) \bar{\Theta}_{n-w-1}. \tag{6.3}$$

Example 6.2. Let π be a spaced seed and $k > 1$. By inserting $(k - 1)$ *'s between every two consecutive positions in π, we obtain a spaced seed π' of the same weight and length $|\pi'| = k|\pi| - k + 1$. It is not hard to see that π' hits the random sequence $R[0, n-1] = s[0]s[1] \cdots s[n-1]$ if and only if π hits one of the following k random sequences:

$$s[i]s[k+i]s[2k+i] \cdots s[lk+i], \qquad i = 0, 1, \ldots, r,$$

$$s[i]s[k+i]s[2k+i]\cdots s[(l-1)k+i], \quad i=r+1, r+2, \ldots, k-1,$$

where $l = \lfloor \frac{n}{k} \rfloor, r = n - kl - 1$. Because π hits the first $r+1$ sequences with probability $\bar{\Pi}_{l+1}$ and the last $k - 1 - r$ sequences with probability $\bar{\Pi}_l$, we have that

$$\bar{\Pi}'_n = (\bar{\Pi}_{l+1})^{r+1} (\bar{\Pi}_l)^{k-1-r}. \tag{6.4}$$

For any $i, j \geq |\pi|$, $\left(\cap_{t=0}^{i-1} \bar{A}_t \right)$ and $\left(\cap_{t=i+|\pi|-1}^{i+j-1} \bar{A}_t \right)$ are independent and $\cap_{t=0}^{i+j-1} \bar{A}_t$ is a subevent of $\left(\cap_0^{i-1} \bar{A}_{t=0} \right) \cap \left(\cap_{t=i+|\pi|-1}^{i+j-1} \bar{A}_t \right)$. Hence,

$$\bar{\Pi}_i \bar{\Pi}_j = \Pr\left[\left(\cap_0^{i-1} \bar{A}_{t=0} \right) \cap \left(\cap_{t=i+|\pi|-1}^{i+j-1} \bar{A}_t \right) \right] > \Pr\left[\cap_{t=0}^{i+j-1} \bar{A}_t \right] = \bar{\Pi}_{i+j}.$$

Hence, formula (6.4) implies that $\bar{\Pi}'_n > \bar{\Pi}_n$ or equivalently $\Pi'_n < \Pi_n$ for any $n \geq |\pi'|$.

6.2.1 A Recurrence System for Hit Probability

We have shown that the non-hit probability of a consecutive seed θ satisfies equation (6.3). Given a consecutive seed θ and $n > |\theta|$, it takes linear-time to compute the hit probability Θ_n. However, calculating the hit probability for an arbitrary seed is rather complicated. In this section, we generalize the recurrence relation (6.3) to a recurrence system in the general case.

For a spaced seed π, we set $m = 2^{|\pi| - w_\pi}$. Let $\mathscr{W}_\pi$ be the set of all m distinct strings obtained from π by filling 0 or 1 in the $*$'s positions. For example, for $\pi = 1 * 11 * 1$,

$$\mathscr{W}_\pi = \{101101, 101111, 111101, 111111\}.$$

The seed π hits the random sequence R at position $n - 1$ if and only if a unique $W_j \in \mathscr{W}_\pi$ occurs at the position. For each j, let $B_n^{(j)}$ denote the event that W_j occurs at the position $n - 1$. Because A_n denotes the event that π hits the sequences R at position $n - 1$, we have that $A_n = \cup_{1 \leq j \leq m} B_n^{(j)}$ and $B_n^{(j)}$'s are disjoint. Setting

$$\pi_n^{(j)} = \Pr\left[\bar{A}_0 \bar{A}_1 \cdots \bar{A}_{n-2} B_{n-1}^{(j)} \right], \quad j = 1, 2, \cdots, m.$$

We have

$$\pi_n = \sum_{1 \leq j \leq m} \pi_n^{(j)}$$

and hence formula (6.2) becomes

$$\bar{\Pi}_n = \bar{\Pi}_{n-1} - \pi_n^{(1)} - \pi_n^{(2)} - \cdots - \pi_n^{(m)}. \tag{6.5}$$

Recall that, for any $W_j \in \mathscr{W}_\pi$ and a, b such that $0 \leq a < b \leq |\pi| - 1$, $W_j[a, b]$ denotes the substring of W_j from position a to position b inclusively. For a string s, we use

$\Pr[s]$ to denote the probability that s occurs at a position $k \geq |s|$. For any i, j, and k such that $1 \leq i, j \leq m$, $1 \leq k \leq |\pi|$, we define

$$p_k^{(ij)} = \begin{cases} \Pr[W_j[k, |\pi| - 1]] & \text{if } W_i[|\pi| - k, |\pi| - 1] = W_j[0, k-1]; \\ 1 & k = |\pi| \ \& \ i = j; \\ 0 & \text{otherwise.} \end{cases}$$

It is easy to see that $p_k^{(ij)}$ is the conditional probability that W_j hits at the position $n+k$ given that W_i hits at position n for $k < |\pi|$ and n.

Theorem 6.1. *Let* $p_j = \Pr[W_j]$ *for* $W_j \in \mathscr{W}_\pi$ $(1 \leq j \leq m)$. *Then, for any* $n \geq |\pi|$,

$$p_j \bar{\Pi}_n = \sum_{i=1}^{m} \sum_{k=1}^{|\pi|} \pi_{n+k}^{(i)} p_k^{(ij)}, \quad j = 1, 2, \ldots, m. \tag{6.6}$$

Proof. For each $1 \leq j \leq m$,

$$p_j \bar{\Pi}_n$$
$$= \Pr\left[\bar{A}_0 \bar{A}_1 \cdots \bar{A}_{n-1} B_{n+|\pi|-1}^{(j)}\right]$$
$$= \sum_{k=1}^{|\pi|-1} \Pr\left[\bar{A}_0 \bar{A}_1 \cdots \bar{A}_{n+k-2} A_{n+k-1} B_{n+|\pi|-1}^{(j)}\right] + \Pr\left[\bar{A}_0 \bar{A}_1 \cdots \bar{A}_{n+|\pi|-2} B_{n+|\pi|-1}^{(j)}\right]$$
$$= \sum_{k=1}^{|\pi|-1} \sum_{i=1}^{m} \Pr\left[\bar{A}_0 \bar{A}_1 \cdots \bar{A}_{n+k-2} B_{n+k-1}^{(i)} B_{n+|\pi|-1}^{(j)}\right] + \pi_{n+|\pi|}^{(j)}$$
$$= \sum_{k=1}^{|\pi|-1} \sum_{i=1}^{m} \Pr\left[\bar{A}_0 \bar{A}_1 \cdots \bar{A}_{n+k-2} B_{n+k-1}^{(i)}\right] \Pr\left[B_{n+|\pi|-1}^{(j)} | B_{n+k-1}^{(i)}\right] + \pi_{n+|\pi|}^{(j)}$$
$$= \sum_{k=1}^{|\pi|-1} \sum_{i=1}^{m} \pi_{n+k}^{(i)} p_k^{(ij)} + \pi_{n+|\pi|}^{(j)}$$
$$= \sum_{i=1}^{m} \sum_{k=1}^{|\pi|} \pi_{n+k}^{(i)} p_k^{(ij)}.$$

This proves formula (6.6).   □

Example 6.3. Let $\pi = 1^a * 1^b$, $a \geq b \geq 1$. Then, $|\pi| = a+b+1$ and $\mathscr{W}_\pi = \{W_1, W_2\} = \{1^a 0 1^b, 1^{a+b+1}\}$. Then we have

$$p_k^{(11)} = p^{|\pi|-k-1} q, \quad k = 1, 2, \ldots, b,$$
$$p_k^{(11)} = 0, \quad k = b+1, b+2, \ldots, |\pi| - 1,$$
$$p_{|\pi|}^{(11)} = 1,$$

$$p_k^{(21)} = p^{|\pi|-k-1}q, \ k = 1,2,\ldots,a,$$
$$p_k^{(21)} = 0, \ k = a+1,a+2,\ldots,|\pi|,$$
$$p_k^{(12)} = p^{|\pi|-k}, \ k = 1,2,\ldots,b,$$
$$p_k^{(12)} = 0, \ k = b+1,b+2,\ldots,|\pi|,$$
$$p_k^{(22)} = p^{|\pi|-k}, \ k = 1,2,\ldots,|\pi|.$$

In addition, $\Pr[W_1] = p^{|\pi|-1}q$ and $\Pr[W_2] = p^{|\pi|}$. Hence we have

$$\begin{cases} p^{|\pi|-1}q\bar{\Pi}_n = \sum_{k=1}^{b} \pi_{n+k}^{(1)}p^{|\pi|-1-k}q + \sum_{k=1}^{a} \pi_{n+k}^{(2)}p^{|\pi|-1-k}q + \pi_{n+|\pi|}^{(1)}, \\ p^{|\pi|}\bar{\Pi}_n = \sum_{k=1}^{b} \pi_{n+k}^{(1)}p^{|\pi|-k} + \sum_{k=1}^{|\pi|} \pi_{n+k}^{(2)}p^{|\pi|-k}. \end{cases}$$

6.2.2 Computing Non-Hit Probability

The recurrence system consisting of (6.5) and (6.6) can be used to calculate the hit probability for arbitrary spaced seeds. Because there are as many as $2^{|\pi|-w_\pi}$ recurrence relations in the system, such a computing method is not very efficient. One may also use a dynamic programming approach for calculating the hit probability. The rationale behind this approach is that the sequences hit by a spaced seed form a regular language and can be represented by a finite automata (or equivalently, a finite directed graph).

Let π be a spaced seed and $n > |\pi|$. For any suffix b of a string in $\mathcal{W}_\pi$ (defined in the last subsection) and $|\pi| \leq i \leq n$, we use $P(i,b)$ to denote the conditional probability that π hits $R[0,i-1]$ given that $R[0,i-1]$ ends with string b, that is,

$$P(i,b) = \Pr\left[\cup_{|\pi|-1\leq j\leq i}A_i \mid R[i-|b|,i-1] = b\right].$$

Clearly, $\Pi(n) = P(n,\varepsilon)$ where ε is the empty string. Furthermore, $P(i,b)$ can be recursively computed as follows:

(i) $P(i,b) = 1$ if $|b| = |\pi|$ and $b \in \mathcal{W}_\pi$;
(ii) $P(i,b) = P(i-1,b \gg 1)$ if $|b| = |\pi|$ and $b \notin \mathcal{W}_\pi$, where $b \gg 1$ denotes the length-$(|b|-1)$ prefix of b; and
(iii) $P(i,b) = (1-p)P(i,0b) + pP(i,1b)$ otherwise.

For most of the spaced seeds, this dynamic programming method is quite efficient. However, its time complexity is still exponential in the worst case. As a matter of fact, computing the hit probability is an NP-hard problem.

6.2.3 Two Inequalities

In this section, we establish two inequalities on hit probability. As we shall see in the following sections, these two inequalities are very useful for comparison of spaced seeds in asymptotic limit.

Theorem 6.2. *Let π be a spaced seed and $n > |\pi|$. Then, for any $2|\pi| - 1 \leq k \leq n$,*
(i) $\pi_k \bar{\Pi}_{n-k+|\pi|-1} \leq \pi_n \leq \pi_k \bar{\Pi}_{n-k}$.
(ii) $\bar{\Pi}_k \bar{\Pi}_{n-k+|\pi|-1} \leq \bar{\Pi}_n < \bar{\Pi}_k \bar{\Pi}_{n-k}$.

Proof. (i). Recall that A_i denotes the event that seed π hits the random sequence R at position $i-1$ and $\bar{A}_i$ the complement of A_i. Set $\bar{A}_{i,j} = \bar{A}_i \bar{A}_{i+1} \ldots \bar{A}_j$. By symmetry,

$$\pi_n = \Pr\left[A_{|\pi|} \bar{A}_{|\pi|+1,n}\right]. \tag{6.7}$$

The second inequality of fact (i) follows directly from that the event $A_{|\pi|} \bar{A}_{|\pi|+1,n}$ is a subevent of $A_{|\pi|} \bar{A}_{|\pi|+1,k} \bar{A}_{k+|\pi|,n}$ for any $|\pi| + 1 < k < n$. The first inequality in fact (i) is proved as follows.

Let k be an integer in the range from $2|\pi| - 1$ to n. For any $1 \leq i \leq |\pi| - 1$, let $\mathscr{S}_i$ be the set of all length-i binary strings. For any $w \in \mathscr{S}_i$, we use E_w to denote the event that $R[k - |\pi| + 2, k - |\pi| + i + 1] = w$ in the random sequence R. With ε being the empty string, E_ε is the whole sample space. Obviously, it follows that $E_\varepsilon = E_0 \cup E_1$. In general, for any string w of length less than $|\pi| - 1$, $E_w = E_{w0} \cup E_{w1}$ and E_{w0} and E_{w1} are disjoint. By conditioning to $A_{|\pi|} \bar{A}_{|\pi|+1,n}$ in formula (6.7), we have

$$\pi_n = \sum_{w \in \mathscr{S}_{|\pi|-1}} \Pr[E_w] \Pr\left[A_{|\pi|} \bar{A}_{|\pi|+1,k} \bar{A}_{k+1,n} | E_w\right]$$

$$= \sum_{w \in \mathscr{S}_{|\pi|-1}} \Pr[E_w] \Pr\left[A_{|\pi|} \bar{A}_{|\pi|+1,k} | E_w\right] \Pr\left[\bar{A}_{k+1,n} | E_w\right]$$

where the last equality follows from the facts: (a) conditioned on E_w, with $w \in \mathscr{S}_{|\pi|-1}$, the event $A_{|\pi|} \bar{A}_{|\pi|+1,k}$ is independent of the positions beyond position k, and (b) $\bar{A}_{k+1,n}$ is independent of the first $k - |\pi| + 1$ positions. Note that

$$\pi_k = \Pr\left[A_{|\pi|} \bar{A}_{|\pi|+1,k}\right] = \Pr\left[A_{|\pi|} \bar{A}_{|\pi|+1,k} | E_\varepsilon\right]$$

and

$$\bar{\Pi}_{n-k+|\pi|-1} = \Pr\left[\bar{A}_{k+1,n} | E_\varepsilon\right].$$

Thus, we only need to prove that

$$\sum_{w \in \mathscr{S}_j} \Pr[E_w] \Pr[A_{|\pi|} \bar{A}_{|\pi|+1,k} | E_w] \Pr[\bar{A}_{k+1,n} | E_w]$$

$$\geq \sum_{w \in \mathscr{S}_{j-1}} \Pr[E_w] \Pr[A_{|\pi|} \bar{A}_{|\pi|+1,k} | E_w] \Pr[\bar{A}_{k+1,n} | E_w] \tag{6.8}$$

for any $1 \leq j \leq |\pi| - 1$ as follows and apply it repeatedly to get the result.

Recall that E_w is the disjoint union of E_{w0} and E_{w1}. By conditioning, we have

$$\Pr\left[A_{|\pi|}\bar{A}_{|\pi|+1,k}|E_w\right] = p\,\Pr\left[A_{|\pi|}\bar{A}_{|\pi|+1,k}|E_{w1}\right] + q\,\Pr\left[A_{|\pi|}\bar{A}_{|\pi|+1,k}|E_{w0}\right]$$

and

$$\Pr\left[\bar{A}_{k+1,n}|E_w\right] = p\Pr\left[\bar{A}_{k+1,n}|E_{w1}\right] + q\Pr\left[\bar{A}_{k+1,n}|E_{w0}\right].$$

Observe that, for any w,

$$\Pr[A_{|\pi|}\bar{A}_{|\pi|+1,k}|E_{w1}] \leq \Pr[A_{|\pi|}\bar{A}_{|\pi|+1,k}|E_{w0}]$$

and

$$\Pr[\bar{A}_{k+1,n}|E_{w1}] \leq \Pr[\bar{A}_{k+1,n}|E_{w0}].$$

By applying Chebyshev's inequality (see the book [87] of Hardy, Littlewood, and Pólya, page 83), we obtain that

$$\Pr\left[A_{|\pi|}\bar{A}_{|\pi|+1,k}|E_w\right]\Pr\left[\bar{A}_{k+1,n}|E_w\right]$$
$$\leq p\Pr\left[A_{|\pi|}\bar{A}_{|\pi|+1,k}|E_{w1}\right]\Pr\left[\bar{A}_{k+1,n}|E_{w1}\right] + q\Pr\left[A_{|\pi|}\bar{A}_{|\pi|+1,k}|E_{w0}\right]\Pr\left[\bar{A}_{k+1,n}|E_{w0}\right].$$

Multiplying $\Pr[E_w]$ on both sides of the last inequality, we obtain

$$\Pr[E_w]\Pr[A_{|\pi|}\bar{A}_{|\pi|+1,k}|E_w]\Pr[\bar{A}_{k+1,n}|E_w]$$
$$\leq \sum_{l=0,1}\Pr[E_{wl}]\Pr[A_{|\pi|}\bar{A}_{|\pi|+1,k}|E_{wl}]\Pr[\bar{A}_{k+1,n}|E_{wl}]$$

from $\Pr[E_{w1}] = \Pr[E_w]p$ and $\Pr[E_{w0}] = \Pr[E_w]q$. Therefore, inequality (6.8) follows from the fact that $\mathscr{S}_j = \{w0, w1 \mid w \in \mathscr{S}_{j-1}\}$. Hence the first inequality in (i) is proved.

(ii). Because $\sum_{j \geq n+1}\pi_j = \bar{\Pi}_n$, the first inequality in (b) follows immediately as follows

$$\bar{\Pi}_k\bar{\Pi}_{n-k+|\pi|-1} = \sum_{j \geq k+1}\pi_j\bar{\Pi}_{n-k+|\pi|-1} \leq \sum_{j \geq k+1}\pi_{n-k+j} = \bar{\Pi}_n.$$

The second inequality in (ii) is similar to its counterpart in (ii) and is obvious. □

6.3 Distance between Non-Overlapping Hits

The following two questions are often asked in the analysis of word patterns: "What is the mean number of times that a pattern hits a random sequence of length N?" and "What is the mean distance between one hit of the pattern and the next?" We call these two questions "number of hits" and "distance between hits." We first study the distance between hits problem. In this section, we use μ_π to denote the expected distance between the non-overlapping hits of a spaced seed π in a random sequence.

6.3.1 A Formula for μ_π

To find the average distance μ_π between non-overlapping hits for a spaced seed π, we define the generating functions

$$U(x) = \sum_{n=0}^{\infty} \bar{\Pi}_n x^n,$$

$$F_i(x) = \sum_{n=0}^{\infty} \pi_n^{(i)} x^n, \ i \leq m.$$

By formula (6.2),

$$\mu_\pi = \sum_{j \geq |\pi|} j\pi_j = |\pi| + \sum_{j \geq |\pi|} \bar{\Pi}_j = U(1) \tag{6.9}$$

where π_j is the first hit probability. Both $U(x)$ and $F_i(x)$ converge when $x \in [0,1]$. Multiplying formula (6.5) by x^{n-1} and summing on n, we obtain

$$(1-x)U(x) + F_1(x) + F_2(x) + \cdots + F_m(x) = 1.$$

Similarly, by formula (6.6), we obtain

$$-x^{|\pi|} p_j U(x) + \sum_{1 \leq i \leq m} C_{ij}(x) F_i(x) = 0, \ 1 \leq j \leq m$$

where $C_{ij}(x) = \sum_{k=1}^{|\pi|} p_k^{(ij)} x^{|\pi|-k}$. Solving the above linear functional equation system, and setting $x = 1$, we obtain the following formula for μ_π.

Theorem 6.3. *Let* $I = [1]_{1 \times m}$ *and* $P = [-p_i]_{m \times 1}$. *Set*

$$A_\pi = [C_{ij}(1)]_{m \times m},$$

and

$$M_\pi = \begin{bmatrix} 0 & I \\ P & A_\pi \end{bmatrix}.$$

Then,

$$\mu_\pi = \det(A_\pi) / \det(M_\pi)$$

where det() *is the determinant of a matrix.*

Example 6.4. For the consecutive seed θ of weight w, $\mathscr{W}_\theta = \{1^w\}$ and $m = 1$. We have

$$C_{11}(x) = \sum_{k=1}^{|\pi|} p^{|\pi|-k} x^{|\pi|-k},$$

$$A_\theta = [\sum_{i=0}^{w-1} p^i],$$

and

$$M_\theta = \begin{bmatrix} 0 & 1 \\ -p^w & \sum_{i=0}^{w-1} p^i \end{bmatrix}.$$

By Theorem 6.3,

$$\mu_\theta = \sum_{i=1}^{w} (1/p)^i.$$

Example 6.5. Continue Example 6.3. For the spaced seed $\pi = 1^a * 1^b$, $a \geq b \geq 1$, we have

$$A_\pi = \begin{bmatrix} \sum_{i=0}^{b-1} p^{a+i}q + 1 & \sum_{i=0}^{a-1} p^{b+i}q \\ \sum_{i=0}^{b-1} p^{a+1+i} & \sum_{i=0}^{a+b} p^i \end{bmatrix}.$$

Therefore,

$$\mu_\pi = \frac{\sum_{i=0}^{a+b} p^i + \sum_{i=0}^{b} \sum_{j=0}^{b-1} p^{a+i+j}q}{p^{a+b}(1 + p(1 - p^b))}.$$

6.3.2 An Upper Bound for μ_π

A spaced seed is *uniform* if its matching positions form an arithmetic sequence. For example, 1**1**1 is uniform with matching position set $\{0, 3, 6\}$ in which the difference between two successive positions is 3. The unique spaced seed of weight 2 and length m is $1 *^{m-2} 1$. Therefore, all the spaced seeds of weight 2 are uniform. In general, a uniform seed is of form $(1*^k)^l 1$, $l \geq 1$ and $k \geq 0$. In Example 6.2, we have showed that $\Pi_i \leq \Theta_i$ for any uniformly spaced seed π and the consecutive seed θ of the same weight. By (6.9), $\mu_\pi \geq \mu_\theta$.

Now we consider non-uniformly spaced seeds. We have proved that $\mu_\theta = \sum_{i=1}^{w_\theta} (1/p)^i$ for consecutive seed θ. For any spaced seed π, by definition, $\mu_\pi \geq |\pi|$. Thus, for any fixed probability p and the consecutive seed θ of the same weight, μ_π can be larger than μ_θ when the length $|\pi|$ of π is large. In this subsection, we shall show that when $|\pi|$ is not too big, μ_π is smaller than μ_θ for any non-uniformly spaced seed π.

For any $0 \leq j \leq |\pi| - 1$, define

$$\mathscr{RP}(\pi) + j = \{i_1 + j, i_2 + j, \cdots, i_{w_\pi} + j\}$$

and let

$$o_\pi(j) = |\mathscr{RP}(\pi) \cap (\mathscr{RP}(\pi) + j)|.$$

Then, $o_\pi(j)$ is the number of 1's that coincide between the seed and the jth shifted version of it. Trivially, $o_\pi(0) = w_\pi$ and $o_\pi(|\pi| - 1) = 1$ hold for any spaced seed π.

Theorem 6.4. *For any spaced seed* π,

$$\mu_\pi \leq \sum_{i=0}^{|\pi|-1} (1/p)^{o_\pi(i)}.$$

Proof. Noticed that the equality holds for the consecutive seeds. Recall that A_j denotes the event that the seed π hits the random sequence at position j and $\bar{A}_j$ the complement of A_j. Let $m_\pi(j) = w_\pi - o_\pi(j)$. We have

$$\Pr\left[A_{n-1} | A_{n-j-1}\right] = p^{m_\pi(j)}$$

for any n and $j \leq |\pi|$. Because A_{n-1} is negatively correlated with the joint event $\bar{A}_0 \bar{A}_1 \cdots \bar{A}_{n-j-2}$,

$$\Pr\left[\bar{A}_0 \bar{A}_1 \cdots \bar{A}_{n-j-2} | A_{n-j-1} A_{n-1}\right] \leq \Pr\left[\bar{A}_0 \bar{A}_1 \cdots \bar{A}_{n-j-2} | A_{n-j-1}\right]$$

for any n and $j \leq |\pi|$. Combining these two formulas together, we have

$$
\begin{aligned}
&\Pr\left[\bar{A}_0 \bar{A}_1 \cdots \bar{A}_{n-j-2} A_{n-j-1} A_{n-1}\right] \\
&= \Pr\left[A_{n-j-1} A_{n-1}\right] \Pr\left[\bar{A}_0 \bar{A}_1 \cdots \bar{A}_{n-j-2} | A_{n-j-1} A_{n-1}\right] \\
&\leq \Pr\left[A_{n-1} | A_{n-j-1}\right] \left\{ \Pr\left[A_{n-j-1}\right] \Pr\left[\bar{A}_0 \bar{A}_1 \cdots \bar{A}_{n-j-2} | A_{n-j-1}\right] \right\} \\
&= p^{m_\pi(j)} \pi_{n-j}
\end{aligned}
$$

Therefore, for any $n \geq |\pi|$,

$$
\begin{aligned}
&\bar{\Pi}_{n-|\pi|} p^{w_\pi} \\
&= \Pr[\bar{A}_0 \bar{A}_1 \cdots \bar{A}_{n-|\pi|-1} A_{n-1}] \\
&= \Pr[\bar{A}_0 \bar{A}_1 \cdots \bar{A}_{n-2} A_{n-1}] + \sum_{i=1}^{|\pi|-1} \Pr[\bar{A}_0 \bar{A}_1 \cdots \bar{A}_{n-|\pi|+i-2} A_{n-|\pi|+i-1} A_{n-1}] \\
&\leq \pi_n + \sum_{i=1}^{|\pi|-1} p^{m_\pi(i)} \pi_{n-|\pi|+i}
\end{aligned}
$$

where we assume $\bar{\Pi}_j = 1$ for $j \leq |\pi| - 1$. Summing the above inequality over n and noting that $\sum_{i=|\pi|}^\infty \pi_i = \Pr[\cup_i A_i] = 1$,

$$\mu_\pi p^{w_\pi} = \sum_{n=0}^\infty \bar{\Pi}_n p^{w_\pi} \leq 1 + \sum_{i=1}^{|\pi|-1} p^{m_\pi(i)} = \sum_{i=0}^{|\pi|-1} p^{m_\pi(i)}$$

or

$$\mu_\pi \leq \sum_{i=0}^{|\pi|-1} p^{m_\pi(i)-w_\pi} = \sum_{i=0}^{|\pi|-1} (1/p)^{o_\pi(i)}.$$

$\square$

Table 6.1 The values of the upper bound in Theorem 6.6 for different w and p after rounding to the nearest integer).

p \ w	10	11	12	13	14
0.6	49	76	121	196	315
0.7	17	21	26	24	44
0.8	11	12	14	15	17
0.9	10	11	12	13	14

Using the above theorem, the following explicit upper bound on μ_π can be proved for non-uniformly spaced seeds π. Its proof is quite involved and so is omitted.

Theorem 6.5. *For any non-uniformly spaced seed π,*

$$\mu_\pi \leq \sum_{i=1}^{w_\pi}(1/p)^i + (|\pi| - w_\pi) - (q/p)\left[(1/p)^{(w_\pi - 2)} - 1\right]$$

6.3.3 Why Do Spaced Seeds Have More Hits?

Recall that, for the consecutive seed θ of weight w, $\mu_\theta = \sum_{i=1}^{w}(\frac{1}{p})^i$. By Theorem 6.5, we have

Theorem 6.6. *Let π be a non-uniformly spaced seed and θ the consecutive seed of the same weight. If $|\pi| < w_\pi + \frac{q}{p}[(\frac{1}{p})^{w_\pi - 2} - 1]$, then, $\mu_\pi < \mu_\theta$.*

Non-overlapping hit of a spaced seed π is a recurrent event with the following convention: If a hit at position i is selected as a non-overlapping hit, then the next non-overlapping hit is the first hit at or after position $i + |\pi|$. By (B.45) in Section B.8, the expected number of the non-overlapping hits of a spaced seed π in a random sequence of length N is approximately $\frac{N}{\mu_\pi}$. Therefore, if $|\pi| < w_\pi + \frac{q}{p}[(\frac{1}{p})^{w_\pi - 2} - 1]$ (see Table 6.1 for the values of this bound for $p = 0.6, 0.7, 0.8, 0.9$ and $10 \leq w \leq 14$), Theorem 6.6 implies that π has on average more non-overlapping hits than θ in a long homologous region with sequence similarity p in the Bernoulli sequence model. Because overlapping hits can only be extended into one local alignment, the above fact indicates that a homology search program with a good spaced seed is usually more sensitive than with the consecutive seed (of the same weight) especially for genome-genome comparison.

6.4 Asymptotic Analysis of Hit Probability

Because of its larger span, in terms of hit probability Π_n, a spaced seed π lags behind the consecutive seed of the same weight at first, but overtakes it when n is relatively big. Therefore, to compare spaced seeds, we should analyze hit probability in asymptotic limit. In this section, we shall show that Π_n is approximately equal to $1 - \alpha_\pi \lambda_\pi^n$ for some α and λ independent of n. Moreover, we also establish a close connection between λ_π and μ_π.

6.4.1 Consecutive Seeds

If a consecutive seed θ does not hit the random sequence R of length n, there must be 0s in the first w_θ positions of R. Hence, by conditioning, we have that $\bar{\Theta}_n$ satisfies the following recurrence relation

$$\bar{\Theta}_n = q\bar{\Theta}_{n-1} + qp\bar{\Theta}_{n-2} + \cdots + qp^{w_\theta-1}\bar{\Theta}_{n-w}, \ n \geq w_\theta, \tag{6.10}$$

where $q = 1 - p$. This recurrence relation has the following characteristic equation

$$f(x) = x^{w_\theta} - q(x^{w_\theta-1} + px^{w_\theta-2} + \cdots + p^{w_\theta-2}x + p^{w_\theta-1}).$$

Let

$$g(x) = f(x)(p - x) = x^{w_\theta}(1 - x) - p^{w_\theta}q.$$

Then,

$$g'(x) = x^{w_\theta-1}[w_\theta - (w_\theta + 1)x].$$

Hence, $g(x)$ and $g'(x)$ have no common factors except for $x - p$ when $p = \frac{w_\theta}{w_\theta+1}$. This implies that $f(x)$ has $w_\theta - 1$ distinct roots for any $0 < p < 1$.

Because $g(x)$ increases in the interval $(0, \frac{w_\theta}{w_\theta+1})$, decreases in the interval $(\frac{w_\theta}{w_\theta+1}, \infty)$, and

$$g(0) = g(1) < 0,$$

$g(x)$ has exactly two positive real roots: p and some r_0. If $p > \frac{w_\theta}{w_\theta+1}$, then, $r_0 \in (0, \frac{w_\theta}{w_\theta+1})$; otherwise, $r_0 \in (\frac{w_\theta}{w_\theta+1}, \infty)$ (see Figure 6.1). Note that r_0 is the unique positive real root of $f(x)$. That $f(r_0) = 0$ implies that

$$\frac{q}{r_0}[1 + \frac{p}{r_0} + \cdots + (\frac{p}{r_0})^{w_\theta-1}] = 1.$$

For any real or complex number r with $|r| \geq r_0$, we have

$$|\frac{q}{r}[1 + \frac{p}{r} + \cdots + (\frac{p}{r})^{w_\theta-1}]| \leq \frac{q}{r_0}[1 + \frac{p}{r_0} + \cdots + (\frac{p}{r_0})^{w_\theta-1}],$$

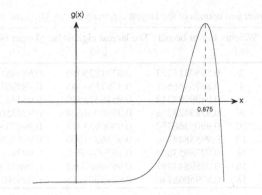

Fig. 6.1 The graph of $g(x) = x^w(1-x) - p^w q$ when w=7 and p=0.7. It increases in $(0, \frac{w}{w+1})$ and decreases in $(\frac{w}{w+1}, 1)$.

where the equality sign is possible only if all terms on the left have the same argument, that is, if $r = r_0$. Hence, r_0 is larger in absolute value than any other root of $f(x)$.

Let $f(x)$ has the following distinct roots

$$r_0, r_1, r_2, \cdots, r_{w_\theta - 1},$$

where $r_0 > |r_1| \geq |r_2| \geq \cdots \geq |r_{w_\theta - 1}|$. Then, By (6.10), we have that

$$\bar{\Theta}_n = a_0 r_0^n + a_1 r_1^n + \cdots + a_{w_\theta - 1} r_{w_\theta - 1}^n. \tag{6.11}$$

where a_is are constants to be determined. Because

$$\theta_i = \bar{\Theta}_{i-1} - \bar{\Theta}_i = p^w q$$

for any $i = w_\theta + 1, \ldots, 2w_\theta$, we obtain the following linear equation system with a_i's as variables

$$\begin{cases} a_0(1-r_0)r_0^{w_\theta} & + \ a_1(1-r_1)r_1^{w_\theta} & + \cdots + a_{w_\theta-1}(1-r_{w_\theta-1})r_{w_\theta-1}^{w_\theta} & = p^w q \\ a_0(1-r_0)r_0^{w_\theta+1} & + \ a_1(1-r_1)r_1^{w_\theta+1} & + \cdots + a_{w_\theta-1}(1-r_{w_\theta-1})r_{w_\theta-1}^{w_\theta+1} & = p^w q \\ & \cdots & & \\ a_0(1-r_0)r_0^{2w_\theta-1} & + \ a_1(1-r_1)r_1^{2w_\theta-1} & + \cdots + a_{w_\theta-1}(1-r_{w_\theta-1})r_{w_\theta-1}^{2w_\theta-1} & = p^w q \end{cases}$$

Solving this linear equation system and using $r_i^{w_\theta}(1-r_i) = p^{w_\theta} q$, we obtain

$$a_i = \frac{p^w q f(1)}{(1-r_i)^2 r_i^{w_\theta} f'(r_i)} = \frac{(p-r_i)r_i}{q[w_\theta - (w_\theta + 1)r_i]}, \quad i = 1, 2, \ldots, w_\theta - 1.$$

Thus, (6.11) implies

Table 6.2 The lower and bounds of the largest eigenvalue r_0 in Theorem 6.7 with $p = 0.70$.

Weight (w)	Lower bound	The largest eigenvalue (r_0)	Upper bound
2	0.5950413223	0.6321825380	0.6844816484
4	0.8675456501	0.8797553586	0.8899014400
6	0.9499988312	0.9528375570	0.9544226080
8	0.9789424218	0.9796064694	0.9798509028
10	0.9905366752	0.9906953413	0.9907330424
12	0.9955848340	0.9956231900	0.9956289742
14	0.9978953884	0.9979046968	0.9979055739
16	0.9989844497	0.9989867082	0.9989868391
18	0.9995065746	0.9995071215	0.9995071407

$$\bar{\Theta}_n = \frac{(p - r_0)}{q[w_\theta - (w_\theta + 1)r_0]} r_0^{n+1} + \varepsilon(n), \tag{6.12}$$

where $\varepsilon(n) = o(r_0^n)$.

The asymptotic formula (6.12) raises two questions:

1. What is the value of r_0?
2. How small is the error term $\varepsilon(n)$ (or the contribution of the $w_\theta - 1$ neglected roots)?

Although we cannot find the exact value of r_0, we have the following bounds on it, whose proof is omitted here.

Theorem 6.7. *Let θ be the consecutive seed of weight w. Then, the r_0 in formula (6.11) is bounded by the following inequalities:*

$$1 - \frac{p^w q}{1 - p^{w+1} - wp^w q} \leq r_0 \leq 1 - \frac{p^w q}{1 - p^{2w} - wp^w q}.$$

Theorem 6.7 gives tight bounds on the largest eigenvalue r_0 for consecutive seeds especially when p is small or n is large. Table 6.2 gives the values of r_0 and the lower and upper bounds for $p = 0.70$ and even weight less than 20.

To analysis the error term, we first prove the following lemma.

Lemma 6.1. *For any $1 \leq i \leq w_\theta - 1$, $|r_i| < p$.*

Proof. We consider the following two cases.

If $p \geq \frac{w_\theta}{w_\theta + 1} > r_0$, the inequality holds trivially.

If $p < \frac{w_\theta}{w_\theta + 1} < r_0$, the equality

$$g(x) = x^{w_\theta + 1}[\frac{1}{x} - 1 - p^{w_\theta} q(\frac{1}{x})^{w_\theta + 1}]$$

implies that all $\frac{1}{r_i}$ and $\frac{1}{p}$ are the roots of the equation

$$y = 1 + p^{w_\theta} q y^{w_\theta + 1}.$$

Let $h(y) = 1 + p^{w_\theta} q y^{w_\theta + 1}$. It intersects the bisector $z = y$ at $\frac{1}{r_0}$ and $\frac{1}{p}$. Moreover, in the interval $[\frac{1}{r_0}, \frac{1}{p}]$, the graph of $z = h(y)$ is convex, and hence the graph of $h(y)$ lies below the bisector. Thus, for any complex number s such that $\frac{1}{r_0} < |s| < \frac{1}{p}$, we have

$$|h(s)| \leq 1 + p^{w_\theta} q |s|^{w_\theta + 1} < |s|.$$

Hence, $s \neq h(s)$ and any complex root of $y = h(y)$ must be larger than $\frac{1}{p}$ in absolute value. This implies that $|r_i| \leq p$. □

For any $1 \leq i \leq w_\theta - 1$, each r_i contributes

$$A_n(r_i) = \frac{p - r_i}{q[w_\theta - (w_\theta + 1)r_i]} r_i^{n+1}$$

to the error term $\varepsilon(n)$. Note that for any complex number s such that $s \leq p < r_0$,

$$\left| \frac{p - s}{w_\theta - (w_\theta + 1)s} \right|$$

reaches its minimum at $s = |s|$ and maximum at $s = -|s|$. We have

$$|A_n(r_i)| \leq \frac{p + |r_i|}{q[w_\theta + (w_\theta + 1)|r_i|]} |r_i|^{n+1} < \frac{2}{q[w_\theta + (w_\theta + 1)p]} p^{n+2}$$

and

$$\varepsilon(n) < \frac{2(w_\theta - 1)}{q[w_\theta + (w_\theta + 1)p]} p^{n+2}.$$

This error estimation indicates that the asymptotic formula (6.12) gives a close approximation to $\bar{\Theta}_n$ and the approximation improves rapidly with n if $p \leq \frac{w_\theta}{w_\theta + 1}$. Table 6.3 gives the exact and approximate values of the non-hit probability for the consecutive seed of weight 3 and $n \leq 10$. When $n = 10$, the approximation is already very accurate.

6.4.2 Spaced Seeds

For an arbitrary spaced seed π, $\bar{\Pi}_n$ does not satisfy a simple recursive relation. To generalize the theorem proved in the last subsection to spaced seeds, we first derive a formula for the generating function $U(z) = \sum_{n=0}^{\infty} \bar{\Pi}_n z^n$ using a "transition matrix." Let

$$\mathcal{V} = \{v_1, v_2, \ldots, v_{2^{|\pi|-1}}\}$$

be the set of all $2^{|\pi|-1}$ binary strings of length $|\pi| - 1$. Define the transition matrix

Table 6.3 The accurate and estimated values of non-hit probability $\bar{\Theta}_n$ with $\theta = 111$ and $p = 0.70$ for $3 \leq n \leq n$.

n	$\bar{\Theta}_n$	From (6.11)	Error
3	0.6570	0.6986	0.0416
4	0.5541	0.5560	0.0019
5	0.4512	0.4426	0.0086
6	0.3483	0.3522	0.0041
7	0.2807	0.2803	0.0004
8	0.2237	0.2231	0.0006
9	0.1772	0.1776	0.0004
10	0.1414	0.1413	0.0001

$$T_\pi = (t_{ij})_{2^{|\pi|-1} \times 2^{|\pi|-1|}}$$

on $\mathcal{V}$ as follows. For any $v_i = v_i[1]v_i[2]\cdots v_i[|\pi|-1]$ and $v_j = v_j[1]v_j[2]\cdots v_j[|\pi|-1]$ in $\mathcal{V}$,

$$t_{ij} = \begin{cases} \Pr[v_j[|\pi|-1]] & \text{if } v_i[2,|\pi|-1] = v_j[1,|\pi|-2] \text{ and } v_i[1]v_j \neq \pi, \\ 0 & \text{otherwise} \end{cases}$$

where $\Pr[v_j[|\pi|-1]] = p$, or q depending on whether $v_j[|\pi|-1] = 1$ or not. For example, if $\pi = 1*1$, then,

$$T_\pi = \begin{pmatrix} q & p & 0 & 0 \\ 0 & 0 & q & p \\ q & 0 & 0 & 0 \\ 0 & 0 & q & 0 \end{pmatrix}.$$

Let $P = (\Pr[v_1], \Pr[v_2], \cdots, \Pr[v_{2^{|\pi|-1}}])$ and $E = (1, 1, \cdots, 1)^t$. By conditional probability, for $n \geq |\pi| - 1$,

$$\bar{\Pi}_n = \sum_{v \in \mathcal{V}} \Pr[v]\Pr[\bar{A}_0\bar{A}_1\cdots\bar{A}_{n-1}|v] = \sum_{i=1}^{2^{|\pi|-1}} \Pr[v_i](\sum_{j=1}^{2^{|\pi|-1}} (T_\pi^{n-|\pi|+1})_{ij}) = PT_\pi^{n-|\pi|+1}E.$$

Hence,

$$U(z)$$
$$= \sum_{i=0}^{|\pi|-2} z^i + \sum_{|\pi|-1}^{\infty} PT_\pi^{n-|\pi|+1}Ez^n$$
$$= \sum_{i=0}^{|\pi|-2} z^i + z^{|\pi|-1} \sum_{n=0}^{\infty} P(zT_\pi)^n E$$

$$= \sum_{i=0}^{|\pi|-2} z^i + z^{|\pi|-1|} \frac{P \mathrm{adj}(I - zT_\pi) E}{\det(I - zT_\pi)}$$

where $\mathrm{adj}(I - zT_\pi)$ and $\det(I - zT_\pi)$ are the adjoint matrix and determinant of $I - zT_\pi$ respectively.

For any $v_i, v_j \in \mathcal{V}$, π does not hit $v_i 0^\pi v_j$. This implies that the power matrix $T_\pi^{2|\pi|-1}$ is positive. Hence, T_π is primitive. By Perron-Frobenius theorem on primitive non-negative matrices, $det(xI - T_\pi)$ has a simple real root $\lambda_1 > 0$ that is larger than the modulus of any other roots. Let $\lambda_2, \cdots, \lambda_{2|\pi|-1}$ be the rest of the roots. Then, $\det(xI - T_\pi) = (x - \lambda_1)(x - \lambda_2) \cdots (x - \lambda_{2|\pi|-1})$, and

$$U(z) = \sum_{i=0}^{|\pi|-2} z^i + z^{|\pi|-1|} \frac{P \mathrm{adj}(I - zT_\pi) E}{(1 - z\lambda_1)(1 - z\lambda_2) \cdots (1 - z\lambda_{2|\pi|-1})}.$$

This implies the following asymptotic expansion result

$$\bar{\Pi}_n = \alpha_1 \lambda_1^n (1 + o(r^n)),$$

where $\alpha_1 > 0$. In summary, we have proved that

Theorem 6.8. *For any spaced seed π, there exists two positive numbers α_π and λ_π such that $\bar{\Pi}_n \approx \alpha_\pi \lambda_\pi^n$*

Theorem 6.9. *For a spaced seed π, λ_π in the above theorem satisfies the following inequalities:*

$$1 - \frac{1}{\mu_\pi - |\pi| + 1} \leq \lambda_\pi \leq 1 - \frac{1}{\mu_\pi}.$$

Proof. For any $n \geq 2|\pi|$ and $k \geq 2$, by the first inequality in Theorem 6.2 (i), $\pi_{n+k} \geq \pi_{n+1} \bar{\Pi}_{|\pi|+k-2}$. Therefore,

$$\frac{\pi_{n+1}}{\bar{Q}_n} = \frac{\pi_{n+1}}{\sum_{i=1}^{\infty} \pi_{n+i}} \leq \frac{\pi_{n+1}}{\pi_{n+1} + \pi_{n+1} \sum_{i=0}^{\infty} \bar{\Pi}_{|\pi|+i}} = \frac{1}{1 + \mu_\pi - |\pi|},$$

and

$$\lambda_\pi = \lim_{n \to \infty} \frac{\bar{\Pi}_{n+1}}{\bar{\Pi}_n} = 1 - \lim_{n \to \infty} \frac{\pi_{n+1}}{\bar{\Pi}_n} \geq 1 - \frac{1}{\mu_\pi - |\pi| + 1}.$$

Similarly, by the second inequality in Theorem 6.2 (i), $\pi_{n+1+j} \leq \pi_{n+1} \bar{\Pi}_j$ for any $j \geq |\pi|$. Therefore,

$$\frac{\pi_{n+1}}{\bar{\Pi}_n} = \frac{\pi_{n+1}}{\sum_{i=1}^{\infty} \pi_{n+i}} \geq \frac{\pi_{n+1}}{\sum_{j=1}^{|\pi|} \pi_{n+j} + \pi_{n+1} \sum_{i=0}^{\infty} \bar{\Pi}_{|\pi|+i}} \geq \frac{\pi_{n+1}}{|\pi| + \sum_{i=0}^{\infty} \bar{\Pi}_{|\pi|+i}} = \frac{1}{\mu_\pi},$$

and

$$\lambda_\pi \leq 1 - \frac{1}{\mu_\pi}.$$

$\square$

Let θ be the consecutive seed of the same weight as π. If π is a uniform seed, we have proved that $\Pi_n < \Theta_n$ for any n in Example 6.2. Moreover, formula (6.4) also implies that $\lambda_\pi = \lambda_\theta$.

If π is non-uniform and $|\pi| < \frac{q}{p}[(\frac{1}{p})^{w_\pi - 2} - 1] + 1$, by Theorems 6.7 and 6.9,

$$\lambda_\pi \leq 1 - \frac{1}{\mu_\pi} < 1 - \frac{1}{\mu_\theta - w_\pi + 1} \leq \lambda_\theta$$

and so $\lim_{n \to \infty} \frac{\Pi_n}{\Theta_n} = 0$. Therefore, there exists a large integer N such that, for any $n \geq N$, $\Pi_n > \Theta_n$. In other words, the spaced seed π will eventually surpass the consecutive seed of the same weight in hit probability if π's length is not too large. This statement raises the following interesting questions:

Question 1 For any non-uniform spaced seed π and the consecutive seed θ of the same weight, is λ_π always smaller than λ_θ?

Question 2 For any non-uniform spaced seed π and the consecutive seed θ of the same weight, does there exist a large integer N such that $\Pi_n > \Theta_n$ whenever $n > N$?

6.5 Spaced Seed Selection

6.5.1 Selection Methods

Consider a homologous region with similarity level p and length n. The *sensitivity* of a spaced seed π on the region largely depends on the hit probability $\Pi_n(p)$. The larger $\Pi_n(p)$ is, the more sensitive π. The spaced seed $\pi = 111*1**1*1**11*111$ was chosen as the default seed in PatternHunter due to the fact that it has the maximum value of $\Pi_{64}(0.70)$ among all the spaced seeds of weight 11.

A straightforward approach to identifying good spaced seeds is through exhaustive search after $\Pi_n(p)$ (for fixed n and p) is computed. The Hedera program designed by Noe and Kocherov takes this approach and is implemented through automata method. Such an approach is quite efficient for short and low-weight seeds. It, however, becomes impractical for designing long seeds as demonstrated by Hedera for two reasons. First, the number of spaced seeds grows rapidly when the number of don't care positions is large. Second, computing hit probability for an arbitrary spaced seed is proved to be NP-hard. Therefore, different heuristic methods are developed for identifying good spaced seeds.

Theorems 6.8 and 6.9 in Section 6.4 suggest that the hit probability of a spaced seed π is closely related to the expected distance μ_π between non-overlapping hits. The smaller the μ_π is, the higher sensitivity the spaced seed π has. Hence, any simple but close approximation of μ_π can be used for the purpose. For example, the upper bound in Theorem 6.4 has been suggested to find good spaced seeds. It is also known that, for any spaced seed π,

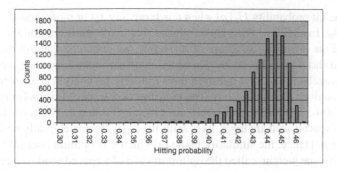

Fig. 6.2 Frequency histogram of spaced seeds of weight 11 and length between 11 and 18, where $n = 64$ and $p = 0.8$. The abscissa interval $[0.30, 0.47]$ is subdivided into 34 bins. Each bar represents the number of spaced seeds whose hit probability falls in the corresponding bin. This is reproduced from the paper [168] of Preparata, Zhang, and Choi; reprinted with permission of Mary Ann Liebert, Inc.

$$\frac{\bar{\Pi}_L}{\pi_L} \leq \mu_\pi \leq \frac{\bar{\Pi}_L}{\pi_L} + |\pi|,$$

where $L = 2|\pi| - 1$. Therefore, $\frac{\bar{\Pi}_L}{\pi_L}$ is another proposed indicator for good spaced seeds.

Sampling method is also useful because the optimal spaced seed is usually unnecessary for practical seed design. As a matter of fact, the optimal spaced seed for one similarity level may not be optimal for another similarity level as we shall see in the next section. Among the spaced seeds with the fixed length and weight, most spaced seeds have nearly optimal hit probability. For instance, Figure 6.2 gives the frequency histogram of all the spaced seeds of weight 11 and length between 11 and 18 when the similarity level set to be 70%. The largest hit probability is 0.467122 and reaches at the default PatternHunter seed. However, more than 86% of the seeds have hit probability 0.430 or more ($> 92\% \times 0.467122$). Therefore, search with hill climbing from a random spaced seed usually finds nearly optimal spaced seed. Mandala program developed by Buhler and Sun takes this approach.

6.5.2 Good Spaced Seeds

Recall that good spaced seeds are selected based on Π_n for a fixed length n and similarity level p of homologous regions. Two natural questions to ask are: Is the best spaced seed with $n = 64$ optimal for every $n > 64$ when p is fixed? and is there a spaced seed π that has the largest hit probability $\Pi_n(p)$ for every $0 < p < 1$ when n is fixed?

To answer the first question, we consider the probabilistic parameters λ_π and μ_π that are studied in Sections 6.3 and 6.4. By Theorems 6.8 and 6.9, λ_π is a dominant factor of the hit probability $\Pi_n(p)$ when n is large, say, 64 or more, and μ_π is closely related to λ_π. In addition, Theorem 6.3 shows that μ_π depends only on the similarity level p and the positional structure of π. Hence, we conclude that the ranking of a spaced seed should be quite stable when the length n of homologous region changes.

Unfortunately, the second question does not have an analytical answer. It is answered through an empirical study. The rankings of top spaced seeds for $n = 64$ and $p = 65\%, 70\%, 75\%, 80\%, 85\%, 90\%$ are given in Table 6.4. When the length and weight of a spaced seed are large, its sensitivity could fluctuate greatly with the similarity level p. Hence, the similarity range in which it is optimal over all the spaced seeds of the same weight, called the optimum span of a spaced seed, is a critical factor when a spaced seed is evaluated. In general, the larger the weight and length of a spaced seed, the narrower its optimum span. Table 6.5 gives the sensitivity Π_{64} for good spaced seeds of weight from 9 to 14 at different similarity levels. The Pattern-Hunter default seed that we mentioned before is the first seed in the row for weight 11 in Table 6.4. It is actually only optimal in the similarity interval $[61\%, 73\%]$ and hence is suitable for detection of remote homologs.

This study also indicates that the spaced seed $111 * 1 * 11 * 1 * *11 * 111$ is probably the best good spaced seed for database search purpose. First, its weight is 12, but its hit probability is larger than the consecutive seed of weight 11 for $n = 64$ and $p = 0.7$. Second, it has rather wide optimum span $[59\%, 96\%]$. Finally, it contains four repeats of $11*$ in its 6-codon span (in the reading frame 5). Such properties are desirable for searching DNA genomic databases in which homologous sequences have diverse similarity and for aligning coding regions.

6.6 Generalizations of Spaced Seeds

6.6.1 Transition Seeds

Substitutions between two pyrimidines (C and T) or between two purines (A and G) are called transitions; other substitutions are called transversions. Studies of mutations in homologous coding sequences indicate that transitions occur approximately three times as frequently as do transversions. This feature is first incorporated into seed design in BLASTZ, where transitions are allowed in just one of the 11 matching positions of a spaced seed. This motivates the study of the transition seeds in which matching and transition positions are fixed and disjoint.

Formally, a transition seed is specified by a string on alphabet $\{1, \#, *\}$, where 1s, #s, and $*$s represent matching positions, transition positions, "don't care" positions, respectively. Consider a transition seed P. Let

$$\mathcal{M} = \{i_1, i_2, \cdots, i_m\}$$

be the set of its matching positions and

Table 6.4 Top-ranked spaced seeds for different similarity levels and weights. The sign '−' means that the corresponding seed is not among the top 10 seeds for the similarity level and weight. This is reproduced from the paper [48] of Choi, Zeng, and Zhang; reprinted with permission of Oxford University Press.

W	Good spaced seeds	Rank under a similarity level (%)					
		65	70	75	80	85	90
9	11*11*1*1***111	1	1	1	1	1	1
	11*1*11***1*111	2	2	2	2	2	3
	11*11**1*1**111	4	4	4	4	4	4
10	11*11***11*1*111	1	1	1	1	1	1
	111**1*1**11*111	2	2	4	6	8	9
	11*11**1*1*1**111	8	6	2	2	2	5
11	111*1**1*1**11*111	1	1	2	2	2	3
	111**1*11**1*1*111	2	2	1	1	1	1
	11*1*1*11**1**1111	6	3	3	5	5	6
12	111*1*11*1**11*111	1	1	1	1	1	1
	111*1**11*1*11*111	2	2	2	5	3	2
	111**1*1*1**11*1111	6	3	3	2	4	4
13	111*1*11**11**1*1111	2	1	1	2	2	2
	111*1**11*1*1**111*111	7	2	2	1	1	1
	111*11*11**1*1*1111	1	4	5	7	8	8
14	111*111**1*11**1*1111	2	1	1	1	1	1
	1111*1**11**11*1*1111	5	2	2	3	3	6
	1111*1*1*11**11*1111	1	3	7	-	-	-
15	1111**1*1*1*11**11*1111	-	5	1	1	1	1
	111*111**1*11**1*11111	-	1	2	5	5	4
	111*111*1*11*1**11111	1	2	-	-	-	-
16	1111*11**11*1*1*11*1111	7	1	2	6	-	-
	1111**11*1*1*11**11*1111	-	7	1	1	1	3
	1111*1**11*1*1**111*1111	1	9	-	-	-	-
	111*111*1**111*11*1111						

$$\mathcal{T} = \{j_1, j_2, \cdots, j_t\}$$

the set of its transition positions. Two sequences S_1 and S_2 exhibit a match of the transition seed P in positions x and y if, for $1 \leq k \leq m$, $S_1[x - L_Q + i_k] = S_2[y - L_Q + i_k]$ and, for $1 \leq k \leq t$, $S_1[x - L_Q + j_k] = S_2[y - L_Q + j_k]$, or two residues $S_1[x - L_Q + j_k]$ and $S_2[y - L_Q + j_k]$ are both pyrimidines or purines.

The analytical studies presented in this chapter can be generalized to transition seeds in a straightforward manner. As a result, good transition seeds can be found using each of the approaches discussed in Section 6.5.1. Hedera and Mandala that were mentioned earlier can be used for transition seed design.

Table 6.5 The hit probability of the best spaced seeds on a region of length 64 for each weight and a similarity level. This is reproduced from the paper [48] of Choi, Zeng, and Zhang; reprinted with permission of Oxford University Press.

W	Best spaced seeds	Similarity (%)	Sensitivity
9	11*11*1*1***111	65	0.52215
		70	0.72916
		80	0.97249
		90	0.99991
10	11*11***11*1*111	65	0.38093
		70	0.59574
		80	0.93685
		90	0.99957
11	111*1**1*1**11*111	65	0.26721
		70	0.46712
	111**1*11**1*1*111	80	0.88240
		90	0.99848
12	111*1*11*1**11*111	65	0.18385
		70	0.35643
		80	0.81206
		90	0.99583
13	111*11*11**1*1*1111	65	0.12327
	111*1*11**11**1*1111	70	0.26475
	111*1**11*1**111*111	80	0.73071
		90	0.99063
14	1111*1*1*11**11*1111	65	0.08179
	111*111**1*11**1*1111	70	0.19351
		80	0.66455
		90	0.98168

Empirical studies show that transition seeds have a good trade-off between sensitivity and specificity for homology search in both coding and non-coding regions.

6.6.2 Multiple Spaced Seeds

The idea of spaced seeds is equivalent to using multiple similar word patterns to increase the sensitivity. Naturally, multiple spaced seeds are employed to optimize the sensitivity. In this case, a set of spaced seeds are selected first; then all the hits generated by these seeds are examined to produce local alignments.

The empirical studies by several research groups show that doubling the number of seeds gets better sensitivity than reducing the single seed by one bit. Moreover, for DNA homology search, the former roughly doubles the number of hits, whereas the latter will increase the number of hits by a factor of four. This implies that using multiple seeds gains not only sensitivity, but also speed.

The formulas in Sections 6.2.1, 6.3.1, and 6.4.2 can be easily generalized to the case of multiple spaced seeds. However, because there is a huge number of combinations of multiple spaced seeds, the greedy approach seems to be the only efficient one for multiple-seed selection. Given the number k of spaced seeds to be selected, the greedy approach finds k spaced seeds in k steps. Let S_{i-1} be the set of spaced seeds selected in the first $i-1$ steps, where $1 \leq i < k$ and we assume $S_0 = \phi$. At the ith iteration step, the greedy approach selects a spaced seed π_i satisfying that $S_{i-1} \cup \{\pi_i\}$ has the largest hit probability.

6.6.3 Vector Seed

Another way to generalize the spaced seed idea is to use a real vector **s** and a threshold $T > 0$ like the seeding strategy used in BLASTP. It identifies a hit at position k in an alignment sequence if and only if the inner product $\mathbf{s} \cdot (a_k, a_{k+1}, \cdots, a_{k+|s|-1})$ is at least T, where $|s|$ denotes the dimension of the vector and a_i is the score of the ith column for each i. The tuple $(\mathbf{s}, T)$ is called a *vector seed*. This framework encodes many other seeding strategies that have been proposed so far. For example, the spaced seed $11***1*111$ corresponds to the vector seed $((1,1,0,0,0,1,0,1,1), 5)$. The seeding strategy for the BLASTP, which requires three consecutive positions with total score greater than 13, corresponds to the vector seed $((1, 1, 1), 13)$. Empirical studies show that using multiple vector seeds is not only effective for DNA sequence alignment, but also for protein sequence alignment.

6.7 Bibliographic Notes and Further Reading

In this section, we briefly summarize the most relevant and useful references on the topics covered in the text.

6.1

Although the filtration technique for string matching has been known for a long time [105], the seed idea is first used in BLAST. In comparison of DNA sequences, BLAST first identifies short exact matches of a fixed length (usually 11 bases) and then extends each seed match to both sides until a drop-off score is reached. By observing that the sensitivities of seed models (of the same weight) vary significantly, Ma, Tromp, and Li proposed to use an optimized spaced seed for achieving the highest sensitivity in PatternHunter [131]. Some other seeding strategies had also been developed at about the same time. WABA developed by Kent employs a simple pattern equivalent to the spaced seed 11*11*11*11*11 to align homologous coding regions [110]. BLAT uses a consecutive seed but allows one or two mismatches to occur in any positions of the seed [109]. Random projection idea, originated in the work of Indyk and Motwani [96], was proposed for improving sensitivity in the work of Buhler [33].

6.2

The mathematical study of spaced seeds is rooted in the classical renewal theory and run statistics. The analysis of the hit probability of the consecutive seeds is even found in the textbook by Feller [68]. The hit probability of multiple string patterns was studied in the papers of Schwager [176] and Guibas and Odlyzko [83]. The interested readers are referred to the books of Balakrishnan and Koutras [22], Deonier, Tavaré, and Waterman [58], and Lothaire [129] for the probability and statistical theory of string pattern matching. Most material presented in this text is not covered in the books that were just mentioned.

The recurrence relation (6.3) in the consecutive seed case had been independently discovered by different researchers (see, for example, [22]). A recurrence relation system for the hit probability of multiple string patterns is presented in [176]. The recurrence system (6.5) and (6.6) for the spaced seeds is given by Choi and Zhang [47]. The dynamic programming algorithm in Section 6.2.2 is due to Keich, Li, Ma, and Tromp [108]. Computing the hit probability was proved to be NP-hard by Ma and Li [130] (see also [125]). Theorem 6.2 is found in the paper of Choi and Zhang [47].

6.3

The mean distance μ between non-overlapping string patterns is traditionally computed through generating function [197, 22]. Theorem 6.3 is due to Kong and Zhang [115] and [214]. Theorem 6.4 is due to Keich et al. [108], which was first proved by using martingale theory. The elementary proof presented in Section 6.3.2 is from [214]. Theorem 6.5 is due to Zhang [214].

6.4

The close approximation formula (6.12) and Lemma 6.1 can be found in the book of Feller [68]. The bounds in Theorems 6.7 and 6.9 are due to Zhang [214]. Theorem 6.8 is due to Buhler, Keich, and Sun [34] (see also [152] and Chapter 7 in [129] for proofs).

6.5

The idea of using a close approximation of μ for identifying good spaced seeds appears in several papers. The close formula $\frac{\Pi_{2|\pi|-1}}{\pi_{2|\pi|-1}}$ is used by Choi and Zhang [47]. The upper bound established in Theorem 6.4 is used by Yang et al. [207] and Kong [115]. The random sampling approach is proposed by Buhler, Keich, and Sun [34].

Sampling approach is also studied in the paper of Preparata, Zhang, and Choi [168]. The spaced seeds reported in Section 6.5.2 are from the paper of Choi, Zeng, and Zhang [48].

6.6

To our best knowledge, the idea of multiple spaced seeds is first used in Pattern-Hunter [131] and is further studied in the papers of Li et al. [123], Brejová, Brown, and Vinař [29], Buhler, Keich, and Sun [34], Csürös and Ma [53], Ilie and Ilie [95], Sun and Buhler [185], and Xu et al. [206].

Transition seeds are studied in the papers of Kucherov, Noè, and Roytberg [118], Noè and Kucherov [153], Schwartz et al. [177], Sun and Buhler [186], Yang and Zhang [208], and Zhou and Florea [216].

The vector seed is proposed in the paper of Brejová, Brown, and Vinař [30] for the purpose of protein alignment. In [31], Brown presents a seeding strategy that has approximately the same sensitivity as BLASTP while keeping five times fewer false positives.

Miscellaneous

For information on applications of the seeding idea to multiple sequence alignment, we refer the reader to the papers of Flannick and Batzoglou [72] and Darling et al. [54]. For information on applications of the seeding idea to other string matching problems, we refer the reader to the papers of Burkhardt and Kärkkäinen [36], Califano and Rigoutsos [37], and Farach-Colton et al. [65].

Chapter 7
Local Alignment Statistics

The understanding of the statistical significance of local sequence alignment has improved greatly since Karlin and Altschul published their seminal work [100] on the distribution of optimal ungapped local alignment scores in 1990. In this chapter, we discuss the local alignment statistics that are incorporated into BLAST and other alignment programs. Our discussion focuses on protein sequences for two reasons. First, the analysis for DNA sequences is theoretically similar to, but easier than, that for protein sequences. Second, protein sequence comparison is more sensitive than that of DNA sequences. Nucleotide bases in a DNA sequence have higher-order dependence due to codon bias and other mechanisms, and hence DNA sequences with normal complexity might encode protein sequences with extremely low complexity. Accordingly, the statistical estimations from DNA sequence comparison are often less reliable than those with proteins.

The statistics of local similarity scores are far more complicated than what we shall discuss in this chapter. Many theoretical problems arising from the general case in which gaps are allowed have yet to be well studied, even though they have been investigated for three decades. Our aim is to present the key ideas in the work of Karlin and Altschul on optimal ungapped local alignment scores and its generalizations to gapped local alignment. Basic formulas used in BLAST are also described.

This chapter is divided into five sections. In Section 7.1, we introduce the extreme value type-I distribution. Such a distribution is fundamental to the study of local similarity scores, with and without gaps.

Section 7.2 presents the Karlin and Altschul statistics of local alignment scores. We first prove that maximal segment scores are accurately described by a geometric-like distribution in asymptotic limit in Sections 7.2.1 and 7.2.3; we introduce the Karlin-Altschul sum statistic in Section 7.2.4. Section 7.2.5 summarizes the corresponding results for optimal ungapped local alignment scores. Finally, we discuss the edge effect issue in Section 7.2.6.

The explicit theory is unknown for the distribution of local similarity scores in the case that gaps are allowed. Hence, most studies in this case are empirical. These studies suggest that the optimal local alignment scores also fit an extreme value type-I distribution for most cases of interest. Section 7.3.1 describes a phase

119

transition phenomena for optimal gapped alignment scores. Section 7.3.2 introduces two methods for estimating the key parameters of the distribution. Section 7.3.3 lists the empirical values of these parameters for BLOSUM and PAM matrices.

In Section 7.4, we describe how P-value and E-value (also called Expect value) are calculated for BLAST database search.

Finally, we conclude the chapter with the bibliographic notes in Section 7.5.

7.1 Introduction

In sequence similarity search, homology relationship is inferred based on P-values or its equivalent E-values. If a local alignment has score s, the P-value gives the probability that a local alignment having score s or greater is found by chance. A P-value of 10^{-5} is often used as a cutoff for BLAST database search. It means that with a collection of random query sequences, only once in a hundred thousand of instances would an alignment with that score or greater occur by chance. The smaller the P-value, the greater the belief that the aligned sequences are homologous. Accordingly, two sequences are reported to be homologous if they are aligned extremely well.

Extremes are rare events that do not happen very often. In the 1950s, Emil Julius Gumbel, a German mathematician, proposed new extreme value distributions. These distributions had quickly grown into the extreme value theory, a branch of statistics, which finds numerous applications in industry. One original distribution proposed by Gumbel is the extreme value type-I distribution, whose distribution function is

$$\Pr[S \geq s] = 1 - \exp(-e^{-\lambda(s-u)}), \tag{7.1}$$

where u and λ are called the location and scale parameters of this distribution, respectively. The distribution defined in (7.1) has probability function

$$f(x) = \lambda \exp(-\lambda(x-u) - e^{-\lambda(x-u)}).$$

Using variable substitution

$$z = e^{-\lambda(x-u)},$$

we obtain its mean and variance as

$$\mu = \lambda \int_{-\infty}^{\infty} x f(x) dx$$
$$= \int_{0}^{\infty} (u - \ln(z)/\lambda) e^{-z} dz$$
$$= u \int_{0}^{\infty} e^{-z} dz - (1/\lambda) \int_{0}^{\infty} \ln(z) e^{-z} dz$$
$$= u + \gamma/\lambda, \tag{7.2}$$

and

$$V = \lambda \int_{-\infty}^{\infty} x^2 f(x)dx - \mu^2$$
$$= \int_0^{\infty} (u - \ln(z)/\lambda)^2 e^{-z}dz - (u + \gamma/\lambda)^2$$
$$= \pi^2 \lambda^2/6, \tag{7.3}$$

where γ is Euler's constant $0.57722\ldots$.

Both theoretical and empirical studies suggest that the distributions of optimal local alignment scores S with or without gaps are accurately described by an extreme value type-I distribution. To given this an intuitive account, we consider a simple alignment problem where the score is 1 for matches and $-\infty$ for mismatches. In this case, the optimal local ungapped alignment occurs between the longest exact matching segments of the sequences, after a mismatch. Assume matches occur with probability p. Then the event of a mismatch followed by k matches has probability $(1-p)p^k$. If two sequences of lengths n and m are aligned, this event occurs at nm possible sites. Hence, the expected number of local alignment with score k or more is

$$a = mn(1-p)p^k.$$

When k is large enough, this event is a rare event. We then model this event by the Poisson distribution with parameter a. Therefore, the probability that there is a local alignment with score k or more is approximately

$$1 - e^{-a} = 1 - \exp\left(-mn(1-p)p^k\right).$$

Hence, the best local alignment score in this simple case has the extreme value type-1 distribution (7.1) with

$$u = \ln\left(mn(1-p)\right)/\ln(1/p)$$

and

$$\lambda = \ln(1/p).$$

In general, to study the distribution of optimal local ungapped alignment scores, we need a model of random sequences. Through this chapter, we assume that the two aligned sequence are made up of residues that are drawn independently, with respective probabilities p_i for different residues i. These probabilities (p_i) define the *background frequency distribution* of the aligned sequences. The score for aligning residues i and j is written s_{ij}. Under the condition that the expected score for aligning two randomly chosen residues is negative, i.e.,

$$E(s_{ij}) = \sum_{i,j} p_i p_j s_{ij} < 0, \tag{7.4}$$

the optimal local ungapped alignment scores are proved to approach an extreme value distribution when the aligned sequences are sufficiently long. Moreover, simple formulas are available for the corresponding parameters λ and u.

Fig. 7.1 The accumulative score of the ungapped alignment in (7.7). The circles denote the ladder positions where the accumulative score is lower than any previously reached ones.

The scale parameter λ is the unique positive number satisfying the following equation (see Theorem B.1 for its existence):

$$\sum_{i,j} p_i p_j e^{\lambda s_{ij}} = 1. \tag{7.5}$$

By (7.5), λ depends on the scoring matrix (s_{ij}) and the background frequencies (p_i). It converts pairwise match scores to a probabilistic distribution $\left(p_i p_j e^{\lambda s_{ij}} \right)$.

The location parameter u is given by

$$u = \ln(Kmn)/\lambda, \tag{7.6}$$

where m and n are the lengths of aligned sequences and $K < 1$. K is considered as a space correcting factor because optimal local alignments cannot locate in all mn possible sites. It is analytically given by a geometrically convergent series, depending only on the (p_i) and (s_{ij}) (see, for example, Karlin and Altschul, 1990, [100]).

7.2 Ungapped Local Alignment Scores

Consider a fixed ungapped alignment between two sequences given in (7.7):

$$
\begin{array}{l}
a\,g\,c\,g\,c\,c\,g\,g\,c\,t\,t\,a\,t\,t\,c\,t\,t\,g\,c\,g\,c\,t\,g\,c\,a\,c\,c\,g \\
|\ |\ \ \ |\ |\ \ |\ |\ |\ \ \ \ \ |\ \ \ \ \ \ |\ \ \ \ |\ |\ \ \ |\ |\ |\ |\ | \\
a\,g\,t\,g\,c\,g\,g\,g\,c\,g\,a\,t\,t\,c\,t\,g\,c\,g\,t\,c\,c\,t\,c\,c\,a\,c\,c\,g
\end{array} \tag{7.7}
$$

We use s_j to denote the score of the aligned pair of residues at position j and consider the accumulative score

$$S_k = s_1 + s_2 + \cdots + s_k, \quad k = 1, 2, \ldots.$$

Starting from the left, the accumulative score S_k is graphically represented in Figure 7.1.

A position j in the alignment is called a *ladder position* if the accumulative score s_j is lower than s_i for any $1 \leq i < j$. In Figure 7.1. the ladder positions are indicated by circles. Consider two consecutive ladder positions a and b, where $a < b$. For a position x between a and b, we define the relative accumulative score at x as

$$R_x = S_x - S_a,$$

which is the difference of the accumulative scores at the positions x and a.

In a random ungapped alignment, the relative accumulative score is a random variable that can be considered as a random walk process (see Section B.7 for background knowledge). For example, if a match scores s and occurs with probability p, then R_x can be considered as the distance from 0 in a random walk that moves right with probability p or left with probability $1 - p$ and stops at -1 on the state set $\{-1, 0, 1, \ldots\}$.

The local alignment between a ladder position to the position where the highest relative accumulative score attains before the next ladder position gives a *maximal-scoring segment* (MSS) in the alignment. Accordingly, alignment statistics are based on the estimation of the following two quantities:

(i) The probability distribution of the maximum value that the corresponding random walk ever achieves before stopping at the absorbing state -1, and
(ii) The mean number of steps before the corresponding random walk first reaches the absorbing state -1.

7.2.1 Maximum Segment Scores

When two protein sequences are aligned, scores other than the simple scores 1 and -1 are used for match and mismatches. These scores are taken from a substitution matrix such as the BLOSUM62 matrix. Because match and mismatches score a range of integer values, the accumulative score performs a complicated random walk. We need to apply the advanced random walk theory to study the distribution of local alignment scores for protein sequences in this section.

Consider a random walk that starts at 0 and whose possible step sizes are

$$-d, -d+1, \ldots, -1, 0, 1, \ldots, c-1, c$$

with respective probabilities

$$p_{-d}, p_{-d+1}, \ldots, p_{-1}, p_0, p_1, \ldots, p_{c-1}, p_c$$

such that

(i) $p_{-d} > 0$, $p_c > 0$, and $p_i \geq 0$, $-d < i < c$, and
(ii) the mean step size $\sum_{j=-d}^{c} j p_j < 0$.

Let X_i denote the score of the aligned pair at the ith position. Then, X_is are iid random variables. Let X be a random variable with the same distribution as X_is. The moment generating function of X is

$$E\left(e^{\theta X}\right) = \sum_{j=-d}^{c} p_j e^{j\theta}.$$

We consider the accumulative scores:

$$S_0 = 0,$$

$$S_j = \sum_{i=1}^{j} X_i, \ j = 1, 2, \ldots,$$

and partition the walk into non-negative excursions between the successive descending ladder points in the path:

$$K_0 = 0,$$

$$K_i = min\left\{k \mid k \geq K_{i-1} + 1, \ S_k < S_{K_{i-1}}\right\}, \ i = 1, 2, \ldots. \tag{7.8}$$

Because the mean step size is negative, the $K_i - K_{i-1}$ are positive integer-valued iid random variables. Define Q_i to be the maximal score attained during the ith excursion between K_{i-1} and K_i, i.e.,

$$Q_i = \max_{K_{i-1} \leq k < K_i} \left(S_k - S_{K_{i-1}}\right), \ i = 1, 2, \ldots. \tag{7.9}$$

The Q_is are non-negative iid random variables. In the rest of this section, we focus on estimating $\Pr[Q_1 \leq x]$ and $E(K_1)$.

Define

$$t^+ = min\{k \mid S_k > 0, \ S_i \leq 0, i = 1, 2, \ldots, k-1\}. \tag{7.10}$$

and set $t^+ = \infty$ if $S_i \leq 0$ for all i. Then t^+ is the stopping time of the first positive accumulative score. We define

$$Z^+ = S_{t^+}, \ t^+ < \infty \tag{7.11}$$

and

$$L(y) = \Pr[0 < Z^+ \leq y]. \tag{7.12}$$

Lemma 7.1. *With notations defined above,*

$$E(K_1) = \exp\left\{\sum_{k=1}^{\infty} \frac{1}{k} \Pr[S_k \geq 0]\right\} \tag{7.13}$$

and

$$1 - E\left(e^{\theta Z^+}; t^+ < \infty\right) = \left(1 - E\left(e^{\theta X}\right)\right) \exp\left\{\sum_{k=1}^{\infty} \frac{1}{k} E\left(e^{\theta S_k}; S_k \leq 0\right)\right\}. \quad (7.14)$$

Proof. The formula (7.13) is symmetric to Formula (7.11) on page 396 of the book [67] of Feller, which is proved for strict ascending ladder points.

Because the mean step size is negative in our case, we have that

$$1 - E\left(e^{\theta Z^+}; t^+ < \infty\right) = \exp\left\{-\sum_{k=1}^{\infty} \frac{1}{k} E\left(e^{\theta S_k}; S_k > 0\right)\right\}. \quad (7.15)$$

by using the same argument as in the proof of a result due to Baxter (Theorem 3.1 in Spitzer (1960)). Because

$$E\left(e^{\theta S_k}; S_k > 0\right) + E\left(e^{\theta S_k}; S_k \leq 0\right) = E\left(e^{\theta S_k}\right) = \left(E\left(e^{\theta X}\right)\right)^k$$

for any k and $\ln(1-y) = -\sum_{i=1}^{\infty} \frac{1}{i} y^i$ for $0 \leq y < 1$, the equation (7.15) becomes

$$1 - E\left(e^{\theta Z^+}; t^+ < \infty\right)$$

$$= \exp\left\{-\sum_{k=1}^{\infty} \frac{1}{k} \left(E\left(e^{\theta X}\right)\right)^k + \sum_{k=1}^{\infty} \frac{1}{k} E\left(e^{\theta S_k}; S_k \leq 0\right)\right\}$$

$$= \exp\left\{\ln\left(1 - E\left(e^{\theta X}\right)\right)\right\} \exp\left\{\sum_{k=1}^{\infty} \frac{1}{k} E\left(e^{\theta S_k}; S_k \leq 0\right)\right\}$$

$$= \left(1 - E\left(e^{\theta X}\right)\right) \exp\left\{\sum_{k=1}^{\infty} \frac{1}{k} E\left(e^{\theta S_k}; S_k \leq 0\right)\right\}.$$

This proves the formula (7.14).   □

Setting $\theta = 0$, the equation (7.15) reduces to

$$1 - \Pr[t^+ < \infty] = \exp\left\{-\sum_{k=1}^{\infty} \frac{1}{k} \Pr[S_k > 0]\right\}. \quad (7.16)$$

If we use λ to denote the unique positive root of the equation $E\left(e^{\theta X}\right) = 1$, i.e.,

$$\sum_{j=-d}^{c} p_j e^{j\lambda} = 1, \quad (7.17)$$

then (7.14) implies

$$1 = E\left(e^{\lambda Z^+}; t^+ < \infty\right) = \sum_{k=1}^{\infty} e^{\lambda k} \Pr[Z^+ = k]. \quad (7.18)$$

Because the mean step size is negative, the distribution function

$$S(y) = \Pr[\max_{k \geq 0} S_k \leq y] \tag{7.19}$$

is finite. Moreover, it satisfies the renewal equation

$$
\begin{aligned}
S(y) &= S(-1) + (S(y) - S(-1)) \\
&= \left(1 - \Pr[t^+ < \infty]\right) + \sum_{k=0}^{y} \Pr[Z^+ = k] S(y - k)
\end{aligned} \tag{7.20}
$$

for any $0 \leq y < \infty$. Define

$$V(y) = (1 - S(y)) e^{\lambda y}. \tag{7.21}$$

$V(y)$ satisfies the renewal equation

$$V(y) = \left(\Pr[t^+ < \infty] - \Pr[Z^+ \leq y]\right) e^{\lambda y} + \sum_{k=0}^{y} e^{\lambda k} \Pr[Z^+ = k] V(y - k). \tag{7.22}$$

By equation (7.18), $e^{\lambda k} \Pr[Z^+ = k]$ is a probability distribution. Using the Renewal Theorem in Section B.8, we have that

$$
\lim_{y \to \infty} V(y) \\
= \frac{\sum_{y=0}^{\infty} e^{\lambda y} \left(\Pr[t^+ < \infty] - \Pr[Z^+ \leq y]\right)}{E\left(Z^+ e^{\lambda Z^+}; t^+ < \infty\right)}.
$$

By (7.18), multiplying the numerator by $e^{\lambda} - 1$ yields $1 - \Pr[t^+ < \infty]$ because the step sizes that have nonzero probability have 1 as their greatest common divisor. Hence,

$$
\lim_{y \to \infty} V(y) \\
= \frac{1 - \Pr[t^+ < \infty]}{\left(e^{\lambda} - 1\right) E\left(Z^+ e^{\lambda Z^+}; t^+ < \infty\right)}.
$$

Recall that $Q_1 = \max_{0 \leq k < K_1} S_k$ and define

$$F(x) = \Pr[Q_1 \leq x], \ 0 \leq x < \infty. \tag{7.23}$$

For any positive integer h, define

$$\sigma(0, h) = \min\{k \mid S_k \notin [0, h]\}.$$

By the definition of $S(y)$ in (7.19) and the law of probabilities, expanding according to the outcome of $\sigma(0,y)$ (either $S_{\sigma(0,h)} > y$ or $S_{\sigma(0,h)} < 0$) yields

$$1 - S(h) = 1 - F(h) + \sum_{k=1}^{\infty} (1 - S(h+k)) \Pr[Q_1 < h \text{ and } S_{\sigma(0,h)} = -k].$$

We define Z^- to be the first negative accumulative score, i.e.,

$$Z^- = S_{t^-}, \quad t^- = \min\{k \mid S_k < 0, \ S_i \geq 0, i = 1, 2, \ldots, k-1\}.$$

Multiplying by $e^{\lambda h}$ throughout the last equation and taking limit $h \to \infty$, we obtain

$$\lim_{h \to \infty} e^{\lambda h} (1 - F(h))$$

$$= \lim_{h \to \infty} V(h) \left(1 - \sum_{k=1}^{\infty} e^{-\lambda k} \Pr[S_{\sigma(0,\infty)} = -k] \right)$$

$$= \frac{(1 - \Pr[t^+ < \infty]) \left(1 - E\left(e^{\lambda Z^-}\right) \right)}{(e^{\lambda} - 1) E\left(Z^+ e^{\lambda Z^+}; t^+ < \infty\right)}. \tag{7.24}$$

from (7.21).

Combining (7.13) and (7.16) gives

$$1 - \Pr[t^+ < \infty] = \frac{\exp\left\{ \sum_{k=1}^{\infty} \frac{1}{k} \Pr[S_k = 0] \right\}}{E(K_1)}.$$

Recall that the derivative of a sum of functions can be calculated as the sum of the derivatives of the individual functions. Differentiating (7.14) with respect to θ and setting $\theta = \lambda$ afterwards shows

$$E\left(Z^+ e^{\lambda Z^+}; t^+ < \infty\right)$$

$$= E\left(X e^{\lambda X}\right) \exp\left\{ \sum_{k=1}^{\infty} \frac{1}{k} E\left(e^{\lambda S_k}; S_k \leq 0\right) \right\}$$

$$= E\left(X e^{\lambda X}\right) \exp\left\{ \sum_{k=1}^{\infty} \frac{1}{k} \left(E\left(e^{\lambda S_k}; S_k < 0\right) + \Pr[S_k = 0] \right) \right\}$$

$$= \frac{E\left(X e^{\lambda X}\right) \exp\left\{ \sum_{k=1}^{\infty} \frac{1}{k} \Pr[S_k = 0] \right\}}{1 - E\left(e^{\lambda Z^-}\right)}.$$

Setting

$$C = \frac{(1 - \Pr[Z^+; t^+ < \infty]) \left(1 - E\left(e^{\lambda Z^-}\right) \right)}{E\left(Z^+ e^{\lambda Z^+}; t^+ < \infty\right)} = \frac{\left(1 - E\left(e^{\lambda Z^-}\right) \right)^2}{E\left(X e^{\lambda X}\right) E(K_1)} \tag{7.25}$$

we have proved the following theorem.

Theorem 7.1. *Assume the step sizes that have nonzero probability do not have non-trivial common divisors. With notations defined above,*

$$\lim_{h\to\infty}(1-F(h))e^{\lambda h} = \frac{C}{e^{\lambda}-1}. \tag{7.26}$$

where λ is defined in (7.17).

Consider a random ungapped alignment $\mathscr{A}$. Q_1 defined for the random walk corresponding to $\mathscr{A}$ is equal to a maximal segment score of it. Hence, $1-F(s)$ denotes the probability of a maximal segment having score s or more in $\mathscr{A}$. By Theorem 7.1, the maximal segment scores have a geometric-like distribution (see Section B.3.3 for definition).

7.2.2 E-value and P-value Estimation

In this subsection, we continue the above discussion to derive formulas for the estimation of the tail distribution of maximal segment scores in ungapped alignment.

Consider the successive excursions with associated local maxima Q_k defined in (7.9). Define

$$M(K_m) = max_{k\le m}Q_k \tag{7.27}$$

and

$$C' = \frac{C}{e^{\lambda}-1}. \tag{7.28}$$

Lemma 7.2. *With notations defined above,*

$$\liminf_{m\to\infty} \Pr\left[M(K_m) \le \frac{\ln m}{\lambda}+y\right] \ge \exp\left\{-C'e^{\lambda-\lambda y}\right\},$$

$$\limsup_{m\to\infty} \Pr\left[M(K_m) \le \frac{\ln m}{\lambda}+y\right] \le \exp\left\{-C'e^{-\lambda y}\right\}.$$

Proof. Because $Q_1,Q_2,\cdots,Q_m$ are iid random variables,

$$\Pr[M(K_m) \le x] = (F(x))^m \tag{7.29}$$

for any $0 < x \le \infty$, where $F(x)$ is the distribution function of Q_1 and is defined in (7.23). For any real number $x < \infty$,

$$1-F(\lceil x\rceil) \le 1-F(x) \le 1-F(\lfloor x\rfloor).$$

Hence,

$$\liminf_{x\to\infty}(1-F(x))e^{\lambda x} \geq \liminf_{x\to\infty}(1-F(\lceil x\rceil))e^{\lambda\lceil x\rceil} = \frac{C}{e^\lambda-1} = C'.$$

On the other hand, because $\limsup_{x\to\infty} x - \lfloor x\rfloor = 1$,

$$\limsup_{x\to\infty}(1-F(x))e^{\lambda x} \leq \lim_{x\to\infty}(1-F(\lfloor x\rfloor))e^{\lfloor x\rfloor}e^{\lambda(x-\lfloor x\rfloor)} \leq \frac{C}{e^\lambda-1}e^\lambda = C'e^\lambda.$$

Replacing x by $\ln(m)/\lambda + y$, (7.29) becomes

$$\Pr[M(K_m) \leq \ln(m)/\lambda + y] = (F(\ln(m)/\lambda+y))^m = \frac{1}{\exp\{-m\ln(F(\ln(m)/\lambda+y))\}}.$$

Using $\lim_{z\to 1}\frac{\ln z}{z-1} = 1$, we obtain

$$\limsup_{m\to\infty}\Pr[M(K_m)\leq \ln(m)/\lambda+y]$$

$$= \limsup_{m\to\infty}\frac{1}{\exp\{-m\ln(F(\ln(m)/\lambda+y))\}}$$

$$= \frac{1}{\liminf_{m\to\infty}\exp\{m(1-F(\ln(m)/\lambda+y))\}}$$

$$\leq \frac{1}{\liminf_{m\to\infty}\exp\{mC'e^{-(\ln(m)+\lambda y)}\}}$$

$$= \exp\{-C'e^{-\lambda y}\}.$$

Similarly,

$$\liminf_{m\to\infty}\Pr[M(K_m)\leq \ln(m)/\lambda+y]$$

$$= \liminf_{m\to\infty}\frac{1}{\exp\{-m\ln(F(\ln(m)/\lambda+y))\}}$$

$$= \frac{1}{\limsup_{m\to\infty}\exp\{m(1-F(\ln(m)/\lambda+y))\}}$$

$$\geq \frac{1}{\limsup_{m\to\infty}\exp\{mC'e^\lambda e^{-(\ln(m)+\lambda y)}\}}$$

$$= \exp\{-C'e^\lambda e^{-\lambda y}\}.$$

Hence, the lemma is proved.   □

Now consider the accumulative score walk W corresponding to an ungapped alignment of length n. Let

$$A = E(K_1), \tag{7.30}$$

which is the mean distance between two successive ladder points in the walk. Then the mean number of ladder points is approximately $\frac{n}{A}$ when n is large (see Section B.8). Ignoring edge effects, we derive the following asymptotic bounds from Lemma 7.2 by setting $y = y' + \frac{\ln A}{\lambda}$ and $m = \frac{n}{A}$:

$$\exp\left\{-\frac{C'e^\lambda}{A}e^{-\lambda y'}\right\} \le \Pr\left[M(n) \le \frac{\ln n}{\lambda} + y'\right] \le \exp\left\{-\frac{C'}{A}e^{-\lambda y'}\right\}. \tag{7.31}$$

Set

$$K = \frac{C'}{A}. \tag{7.32}$$

Replacing y' by $(\ln(K) + s)/\lambda$, inequality (7.31) becomes

$$\exp\left\{-e^{\lambda - s}\right\} \le \Pr\left[M(n) \le \ln(Kn) + s/\lambda\right] \le \exp\left\{-e^{-s}\right\},$$

or equivalently

$$\exp\left\{-e^{\lambda - s}\right\} \le \Pr\left[\lambda M(n) - \ln(Kn) \le s\right] \le \exp\left\{-e^{-s}\right\}. \tag{7.33}$$

In the BLAST theory, the expression

$$Y(n) = \lambda M(n) - \ln(Kn)$$

is called the *normalized score* of the alignment. Hence, the P-value corresponding to an observed value s of the normalized score is

$$\text{P-value} \approx 1 - \exp\left\{-e^{-s}\right\}. \tag{7.34}$$

7.2.3 The Number of High-Scoring Segments

By Theorem 7.1 and (7.28), the probability that any maximal-scoring segment has score s or more is approximately $C'e^{-\lambda s}$. By (7.30), to a close approximation there are N/A maximal-scoring segments in a fixed alignment of N columns as discussed in Section 7.2.2. Hence, the expected number of the maximal-scoring segments with score s or more is approximately

$$\frac{NC'}{A}e^{-\lambda s} = NKe^{-\lambda s}, \tag{7.35}$$

7.2.4 Karlin-Altschul Sum Statistic

When two sequences are aligned, insertions and deletions can break a long alignment into several parts. If this is the case, focusing on the single highest-scoring
segment could lose useful information. As an option, one may consider the scores
of the multiple highest-scoring segments.

Denote the r disjoint highest segment scores as

$$M^{(1)} = M(n), M^{(2)}, \ldots, M^{(r)}$$

in an alignment with n columns. We consider the normalized scores

$$S^{(i)} = \lambda M^{(i)} - \ln(Kn), \quad i = 1, 2, \ldots, r. \tag{7.36}$$

Karlin and Altschul (1993) showed that the limiting joint density $f_S(x_1, x_2, \cdots, x_r)$
of $S = (S^{(1)}, S^{(2)}, \cdots, S^{(r)})$ is

$$f_S(x_1, x_2, \cdots, x_r) = \exp\left(-e^{-x_r} - \sum_{k=1}^{r} x_k\right) \tag{7.37}$$

in the domain $x_1 \geq x_2 \geq \ldots \geq x_r$.

Assessing multiple highest-scoring segments is more involved than it might first
appear. Suppose, for example, comparison X reports two highest scores 108 and 88,
whereas comparison Y reports 99 and 90. One can say that Y is not better than X,
because its high score is lower than that of X. But neither is X considered better,
because the second high score of X is lower than that of Y. The natural way to
rank all the possible results is to consider the sum of the normalized scores of the r
highest-scoring segments

$$S_{n,r} = S^{(1)} + S^{(2)} + \cdots + S^{(r)} \tag{7.38}$$

as suggested by Karlin and Altschul. This sum is now called the *Karlin-Altschul
sum statistic*.

Theorem 7.2. *The limiting density function of Karlin-Altschul sum $S_{n,r}$ is*

$$f_{n,r}(x) = \frac{e^{-x}}{r!(r-2)!} \int_0^{\infty} y^{r-2} \exp\left(-e^{(y-x)/r}\right) dy. \tag{7.39}$$

Integrating $f_{n,r}(x)$ from t to ∞ gives the tail probability that $S_{n,r} \geq t$. This resulting double integral can be easily calculated numerically. Asymptotically, the tail

probability is well approximated by the formula

$$\Pr[S_{n,r} \geq t] = \frac{e^{-t}t^{r-1}}{r!(r-1)!}. \tag{7.40}$$

especially when $t > r(r-1)$. This is the formula used in the BLAST program for calculating p-value when multiple highest-scoring segments are reported.

7.2.5 Local Ungapped Alignment

The statistic theory of Sections 7.2.1 and 7.2.2 considered the maximal scoring segments in a fixed ungapped alignment. In practice, the objective of database search is to find all good matches between a query sequence and the sequences in a database. Here, we consider a general problem of calculating the statistical significance of a local ungapped alignment between two sequences. The sequences in a highest-scoring local ungapped alignment between two sequences is called the *maximal-scoring segment pairs* (MSP).

Consider two sequences of length n_1 and n_2. To find the MSPs of the sequences, we have to consider all $n_1 + n_2 - 1$ possible ungapped alignments between the sequences. Each such alignment yields a random work as that studied in Section 7.2.1. Because the $n_1 + n_2 - 1$ corresponding random walks are not independent, it is much more involved to estimate the mean value of the maximum segment score. The theory developed by Dembo et al. [56, 57] for this general case is too advanced to be discussed in this book. Here we simply state the relevant results.

To some extent, the key formulas in Sections 7.2.2 can be taken over to the general case, with n being simply replaced by $n_1 n_2$. Consider the sequences $x_1 x_2 \ldots x_{n_1}$ and $y_1 y_2 \ldots y_{n_2}$, where x_i and y_j are residues. We use $s(x_i, y_j)$ to denote the score for aligning residues x_i and y_j. The optimal local ungapped alignment score S_{max} from the comparison of the sequences is

$$S_{max} = \max_{\Delta \leq \min\{n_1, n_2\}} \max_{\substack{i \leq n_1 - \Delta \\ j \leq n_2 - \Delta}} \sum_{l=1}^{\Delta} s(x_{i+l}, x_{j+l}).$$

Suppose the sequences are random and independent: x_i and y_j follows the same distribution. The random variable S_{max} has the following tail probability distribution:

$$\Pr[S_{max} > \frac{1}{\lambda} \log(n_1 n_2) + y] \approx 1 - e^{-Ke^{-\lambda y}}, \tag{7.41}$$

where λ is given by equation (7.5) and K is given by equation (7.32) with n replaced by $n_1 n_2$. The mean value of S_{max} is approximately

$$\frac{1}{\lambda} \log(K n_1 n_2). \tag{7.42}$$

If the definition of the normalized score S is modified as

$$S = \lambda S_{max} - \log(Kn_1n_2),\qquad(7.43)$$

the P-value corresponding to an observed value s of S is

$$\text{P-value} \approx 1 - e^{-(e^{-s})}.\qquad(7.44)$$

The expected number of the MSPs scoring s or more is

$$Kn_1n_2e^{\lambda s}\qquad(7.45)$$

The above theory is developed under several conditions and only applies, for example, when the sequences are sufficiently long and the aligned sequences grow at similar rate. But empirical studies show that the above theory carries over essentially unchanged after edge effect is made in practical cases in which the aligned sequences are only a few hundred of base pairs long.

7.2.6 Edge Effects

When the aligned sequences have finite lengths n_1 and n_2, optimal local ungapped alignment will tend not to appear at the end of a sequence. As a result, the optimal local alignment score will be less than that predicated by theory. Therefore, edge effects have to be taken into account by subtracting from n_1 and n_2 the mean length of MSPs.

Let $E(L)$ denote the mean length of a MSP. Then effective lengths for the sequence compared are

$$n_i' = n_i - E(L),\ \ i = 1,2.\qquad(7.46)$$

The normalized score (7.43) becomes

$$S' = \lambda S_{max} - \log(Kn_1'n_2').\qquad(7.47)$$

The expected number of MSPs scoring s or more given in (7.45) is adjusted as

$$Kn_1'n_2'e^{-\lambda s}.\qquad(7.48)$$

Given that the score of a MSP is denoted by s, the mean length $E(L)$ of this MSP is obtained from dividing s by the expected score of aligning a pair of residues:

$$E(L) = \frac{s}{\sum_{ij} q_{ij}s_{ij}},\qquad(7.49)$$

where q_{ij} is the target frequency at which we expect to see residue i aligned with residue j in the MSP, and s_{ij} is the score for aligning i and j. With the value of λ, the s_{ij} and the background frequencies q_i and q_j in hand, q_{ij} can be calculated as

$$q_{ij} \approx p_i p_j e^{\lambda s_{ij}}. \tag{7.50}$$

Simulation shows that the values calculated from (7.49) are often larger than the empirical mean lengths of MSPs especially when n_1 and n_2 are in the range from 10^2 to 10^3 (Altschul and Gish, 1996, [6]). Accordingly, the effective lengths defined by (7.46) might lead to P-value estimates less than the correct values. The current version of BLAST calculates empirically the mean length of MSPs in database search.

7.3 Gapped Local Alignment Scores

In this section, we concern with the optimal local alignment scores S (in the general case that gaps are allowed). Although the explicit theory is unknown in this case, a number of empirical studies strongly suggest that S also has asymptotically an extreme value distribution (7.1) under certain conditions on the scoring matrix and gap penalty used for alignment. As we will see, these conditions are satisfied for most combinations of scoring matrices and gap penalty costs.

7.3.1 Effects of Gap Penalty

Most empirical studies focus on the statistical distribution of the scores of optimal local alignments with affine gap costs. Each such gap cost has gap opening penalty o and gap extension penalty e, by which a gap of length k receives a score of $-(o + k \times e)$.

Consider two sequences X' and X'' that are random with letters generated independently according to a probabilistic distribution. The optimal local alignment score S_{max} of X' and X'' depends on the sequence lengths m and n, and the letter distribution, the substitution matrix, and affine gap cost (o, e). Although the exact probabilistic distribution of S_{max} is unclear, S_{max} has either linear or logarithmic growth as m and n go to infinite.

This phase transition phenomenon for the optimal local alignment score was studied by Arratia and Waterman [15]. Although a rigorous treatment is far beyond the scope of this book, an intuitive account is quite straightforward. Consider two sequences $x_1 x_2 \ldots x_n$ and $y_1 y_2 \ldots y_n$. Let $S(t)$ denote the score of the optimal alignment of $x_1 x_2 \ldots x_t$ and $y_1 y_2 \ldots y_t$. Then,

$$S_{t+k} \geq S_t + S(x_{t+1} x_{t+2} \cdots x_{t+k}, y_{t+1} y_{t+2} \cdots y_{t+k}).$$

Table 7.1 The average amino acid frequencies reported by Robinson and Robinson in [173].

Amino acid	Freqency	Amino acid	Freqency	Amino acid	Freqency	Amino acid	Freqency
Ala	0.078	Gln	0.043	Leu	0.090	Ser	0.071
Arg	0.051	Glu	0.063	Lys	0.057	Thr	0.058
Asn	0.045	Gly	0.074	Met	0.022	Trp	0.013
Asp	0.054	His	0.022	Phe	0.039	Tyr	0.032
Cys	0.019	Ile	0.051	Pro	0.052	Val	0.064

Because $S(x_{t+1}x_{t+2}\cdots x_{t+k}, y_{t+1}y_{t+2}\cdots y_{t+k})$ equals S_k in distribution, S_t and hence $E(S_t)$ satisfy the subadditive property. This implies that the following limit exists and equals the supremum

$$\lim \frac{E(S_t)}{t} = \sup_{t \geq 1} \frac{E(S_t)}{t} = c,$$

where c is a constant that depends only on the penalty parameters and the letter distribution.

When the penalty cost is small, the expected score of each aligned pair is positive; then the limit c is positive. In this case, the score S_t grows linearly in t. For example, if the score is 1 for matches and 0 otherwise, the optimal local alignment score is equal to the length of the longest subsequence common to the sequences. The alignment score converges to $c \log n$ for some $c \in (1/k, 1)$ as n goes to infinity if the aligned sequences have the same length n and are generated by drawing uniformly k letters.

When the penalty cost is large such that the expected score of each aligned pair is negative, the optimal local alignments, having positive score, represent large deviation behavior. In this case, $c = 0$ and the probability that a local alignment has a positive score decays exponentially fast in its length. Hence, S_t grows like $\log(t)$. The region consisting of such penalty parameters is called the logarithmic region. In this region, local alignments of positive scores are rare events. By using the Chen-Stein method (see, for example, [44]), Poisson approximation can be established to show that the optimal local alignment score approaches an extreme value distribution; see, for example, the papers of Karlin and Altschul [100] and Arratia, Gordan, and Waterman [11]. Empirical studies demonstrate that affine gap costs (o, e) satisfying $o \geq 9$ and $e \geq 1$ are in the logarithmic region if the BLOSUM62 matrix is used.

7.3.2 Estimation of Statistical Parameters

There are no formulas for calculating the relevant parameters for the hypothetical extreme value distribution of the optimal local alignment scores. These parameters

Table 7.2 Empirical values for l, u, λ, and K. Data are from Methods in Enzymology, Vol. 26, Altschul and Gish, Local alignment statistics, 460-680, Copyright (1996), with permission from Elsevier [6].

Sequence length n	Mean alignment length l	u	λ	K
403	32.4	32.04	0.275	0.041
518	36.3	33.92	0.279	0.048
854	43.9	37.84	0.272	0.040
1408	51.6	41.71	0.268	0.036
1808	55.1	43.54	0.271	0.041
2322	59.1	45.53	0.267	0.035
2981	63.5	47.32	0.270	0.040

have to be estimated through simulation for measuring statistical significance of local alignments.

The most straightforward way is to generate the optimal local alignment scores of random protein sequences and estimate the parameters u and λ by fitting them to these scores.

Taking this "direct" approach, Altschul and Gish generated 10,000 pairs of random sequences with the average amino acid frequencies listed in Table 7.1 for each of a large set of sequence lengths [6]. For each pair of sequences of the same length, the optimal score, together with the length of each optimal local alignment, is calculated with the BLOSUM62 matrix and gap penalty cost (12, 1).

Altschul and Gish estimated u and λ by fitting them to the data using the method of moments. The parameters u and λ are calculated from the sample mean and variance using the formulas (7.2) and (7.3). Some empirical values for μ and λ are shown in Table 7.2.

Rather than using the optimal local alignment scores for pairs of random sequences, one may use maximal local alignment scores to estimate λ and K. The Smith-Waterman algorithm for aligning locally two sequences generates a score s_{ij} for each cell (i, j) in the alignment graph. It is the score of the best local alignment ending at the cell (i, j). This alignment corresponds uniquely to a path from some cell (i', j') to (i, j) in the alignment graph, where $i' \le i$ and $j' \le j$. An island with anchor (i', j') consists of all cells on the paths starting at (i', j') and corresponding to best local alignments. The score assigned to an island is the maximum score of the cells it contains. By modifying the standard implementation of the Smith-Waterman algorithm, one can obtain all island scores generated in a tabular computation with a little extra computation.

Island scores are the scores of distinct optimal local alignments. Thus, the number of islands with score s or more satisfies equation (7.45). Because the event of an island score being s or more has a geometric-like distribution, one may use a simple maximum likelihood method to estimate λ (see the gray box below).

Let I_s denote the set of islands having score s or more. Then, the average score in excess of s of these islands is

$$A(s) = \frac{1}{|I_s|} \sum_{i \in I_s} (S(i) - s),$$ (7.51)

where $S(i)$ is the score of island i and $|I_s|$ the number of islands in I_s. Because the island scores are integral and have no proper common divisors in the cases of interest, the maximum-likelihood estimate for λ is

$$\lambda_s = \ln(1 + \frac{1}{A(s)}).$$ (7.52)

By equation (7.45), K is calculated as

$$K_s = \frac{1}{V} \times |I_s| \times e^{s\lambda_s},$$ (7.53)

where V is the size of the search space from which the island scores are collected. V is equal to n^2 if the islands are obtained from aligning locally two sequences of length n; it is Nn^2 if N such alignments were performed.

Maximum-Likelihood Method for Estimating λ (Altschul et al., 2001, [5])

The island scores S follow asymptotically a geometric-like distribution

$$\Pr[S = x] = De^{-\lambda x},$$

where D is a constant. For a large integer cutoff c,

$$\Pr[S = x | S \geq c] \approx \frac{De^{-\lambda x}}{\sum_{j=c}^{\infty} De^{-\lambda j}} = (1 - e^{-\lambda})e^{-\lambda(x-c)}.$$

Let x_i denote the ith island scores for $i = 1, 2, \ldots, M$. Then the logarithm of the probability that all x_is have a value of c or greater is

$$\ln(\Pr[x_1, x_2, \ldots, x_M | x_1 \geq c, x_2 \geq c, \ldots, x_M \geq c])$$
$$= -\lambda \sum_{j=1}^{M} (x_j - c) + M \ln(1 - e^{-\lambda}).$$

The best value λ_{ML} of λ is the one that maximizes this expression. By equating the first derivation of this expression to zero, we obtain that

$$\lambda_{ML} = \ln \left(1 + \frac{1}{\frac{1}{M} \sum_{j=1}^{M} (x_j - c)} \right).$$

Because we obtain the estimates of λ and K from finite-length sequences, edge effect has to be corrected. This can be done by recording only islands within the central region of size $(n - l) \times (n - l)$ for some chosen integer l. For example, we can choose l to be the mean length of the best local alignment.

Another key issue for applying the island method effectively is how to choose an appropriate cutoff value s. Notice that optimal local alignments having low score likely contains no gaps. When s is too small, the estimate λ_{ML} of λ is biased toward the higher value that is for optimal local ungapped alignment scores. On the other hand, when s is too large, there are fewer islands having score s or greater and so the standard error of λ_{ML} becomes larger.

We have introduced two methods for estimating statistical parameters λ and K. The island method has a significant speed advantage especially for long sequences. The primary reason for this advantage is that for each comparison of two sequences, the island method generates multiple data points.

7.3.3 Statistical Parameters for BLOSUM and PAM Matrices

For practical purpose, the parameters λ, K, together with relative entropy H, were empirically determined by Altschul and Gish for popular scoring matrices. The parameters for the BLOSUM62 and PAM250 matrices are compiled in Table 7.3. The numbers for infinite gap costs were derived from theory; others were obtained based on the average amino acid frequencies listed in Table 7.1. They may be poor for other protein sequence models.

From the data presented in Table 7.3, Altschul and Gish (1996, [6]) observed two interesting facts. First, for a given gap opening cost o, the parameters λ, u, and K remain the same when gap extension costs e are relatively large. It implies that, with these affine gap costs, optimal local alignments that occur by chance do not likely contain insertion and deletions that involve more than 1 residue. Hence, for a given o, it is unrewarding to use any gap extension cost that is close to o.

Second, the ratio of λ to λ_{∞} for the ungapped case indicates the proportion of information in local ungapped alignments is lost in the hope of extending the alignments using gaps. Low gap costs are sometimes used in hoping that local alignments containing long insertions or deletions will not be severely penalized. But, this decreases the information contained in each aligned pair of residues. As an alternative method, one may employ fairly high gap costs and evaluate the multiple high-scoring local alignments using the Karlin-Altschul sum statistic presented in Section 7.2.4.

Table 7.3 Parameters λ, K, and H for PAM250 (left column) and BLOSUM62 (right column) in conjunction with affine gap costs (o, e) and the average amino acid frequencies listed in Table 7.1. Here the relative entropy is calculated using the natural logarithm. Data are from Methods in Enzymology, Vol. 26, Altschul and Gish, Local alignment statistics, 460-680, Copyright (1996), with permission from Elsevier [6].

o	e	λ	K	H	o	e	λ	K	H
∞		0.229	0.090	0.23	∞		0.318	0.130	0.40
16	4-16	0.217	0.070	0.21	12	3-12	0.305	0.100	0.38
	3	0.208	0.050	0.18		2	0.300	0.009	0.34
	2	0.200	0.040	0.16		1	0.275	0.050	0.25
	1	0.172	0.018	0.09					
					11	3-11	0.301	0.09	0.36
14	6-14	0.212	0.060	0.19		2	0.286	0.07	0.29
	4, 5	0.204	0.050	0.17		1	0.255	0.035	0.19
	3	0.194	0.035	0.14					
	2	0.180	0.025	0.11	10	4-10	0.293	0.080	0.33
	1	0.131	0.008	0.04		3	0.281	0.060	0.29
						2	0.266	0.040	0.24
12	7-12	0.199	0.050	0.15		1	0.216	0.014	0.12
	5,6	0.191	0.040	0.13					
	4	0.181	0.029	0.12	9	5-9	0.286	0.080	0.29
	3	0.170	0.022	0.10		3,4	0.273	0.060	0.25
	2	0.145	0.012	0.06		2	0.244	0.030	0.18
						1	0.176	0.008	0.06
10	8-10	0.175	0.031	0.11					
	7	0.171	0.029	0.10	7	6-7	0.247	0.050	0.18
	6	0.165	0.024	0.09		4,5	0.230	0.030	0.15
	5	0.158	0.020	0.08		3	0.208	0.021	0.11
	4	0.148	0.017	0.07		2	0.164	0.009	0.06
	3	0.129	0.012	0.05					
					6	5,6	0.200	0.021	0.10
8	7,8	0.123	0.014	0.05		4	0.179	0.014	0.08
	6	0.115	0.012	0.04		3	0.153	0.010	0.05
	5	0.107	0.011	0.03					

7.4 BLAST Database Search

We now consider the most relevant case in practice. In this case, we have a query sequence and a database, and we wish to search the entire database to find all sequences that are homologies of the query sequence. In this case, significant high-scoring local alignments, together with their P-values and E-values, are reported. In this section, calculations of P-values and E-values in BLAST are discussed.

7.4.1 Calculation of P-values and E-values

Database search is actually a multiple testing problem as a database contains many sequences. P-value and E-value calculations must be amended to account for the size of the database.

We first consider the case that distinct high-scoring alignments are listed. Assume that we obtain a high-scoring alignment of score s. Let l_Q be the length of the query sequence Q. Consider an individual sequence T of length l_T in the database $\mathscr{D}$. The mean number E_T of local alignments with score s or more occurring in the comparison of Q and T is calculated from (7.48) as

$$E_T = K(l_Q - \bar{l}(s))(l_T - \bar{l}(s))e^{-\lambda s}, \qquad (7.54)$$

where $\bar{l}(s)$ is the length adjustment, the mean length of high-scoring alignments with score s. Let N be the number of sequences in the database. Then, the effective size of search space is

$$\text{eff-searchSP} = \sum_{T \in \mathscr{D}} l_T - N\bar{l}(s). \qquad (7.55)$$

By the linearity property of means, the expected number of high-scoring alignments with score s or more found in the entire database becomes

$$\text{Expect} = \sum_{T \in \mathscr{D}} E_T = K \times (l_Q - \bar{l}(s)) \times \text{eff-searchSP} \times e^{-\lambda s}. \qquad (7.56)$$

The P-value can be calculated from Expect as

$$\text{P-value} \approx 1 - e^{-\text{Expect}}. \qquad (7.57)$$

For certain types of search, BLAST evaluates the statistical significance of a high-scoring alignment based on Karlin-Altschul sum statistic. In the earlier version of BLAST, it was done only for ungapped alignments as an alternative to performing gapped alignment. The sum P-value is only calculated for an admissible set of high-scoring alignments. The P-value reported for any alignment is the smallest of the sum P-values obtained with the admissible sets that contain the alignment.

Let $\{A_{i_j}\}$ be a subset of alignments, sorted in ascending order by the offset in the query sequence. It is admissible if there are three fixed constants d, d_q, and d_s such that, for any adjacent alignments A_{i_k} and $A_{i_{k+1}}$, their offset and end positions in the query sequence satisfy

$$\text{end}(A_{i_k}) \leq \text{end}(A_{i_{k+1}}),$$
$$\text{offset}(A_{i_k}) \leq \text{offset}(A_{i_{k+1}}),$$
$$\text{end}(A_{i_k}) - d \leq \text{offset}(A_{i_{k+1}}) \leq \text{end}(A_{i_k}) + d_q,$$

and their offset and end positions in the database sequence satisfy the same inequalities with d_q replaced by d_s.

For an admissible set $\mathscr{A}$ of alignments, its sum P-value and E-value are calculated using the sum of the normalized scores of the alignments

$$S(\mathscr{A}) = \sum_{A \in \mathscr{A}} (\lambda_A S_A - \ln(K_A)), \tag{7.58}$$

where S_A is the raw score of A and λ_A and K_A are the parameters associated with A. In most cases, these statistical parameters in (7.58) have the same value. However, they can be different when BLASTX is used.

Assume that $\mathscr{A}$ contains r alignments and the query and subject sequences have length m and n respectively. $S(\mathscr{A})$ is further adjusted using r, m, and n as

$$S'(\mathscr{A}) = S(\mathscr{A}) - \ln(nm) - (r-1)\ln(d_q d_s) - \ln(r!). \tag{7.59}$$

The sum P-value is then calculated from (7.39) by integration:

$$\text{P-value} = \frac{r^{r-2}}{(r-1)!(r-2)!} \int_{S'(\mathscr{A})}^{\infty} e^{-x} \int_0^{\infty} y^{r-2} exp\left(-e^{(y-x)/r}\right) dydx. \tag{7.60}$$

E-value is then calculated by multiplying a factor of eff-searchSP/mn

$$\text{Expect}_{\mathscr{A}} = -\ln(1 - \text{P-value}) \times \frac{\text{eff-searchSP}}{mn}, \tag{7.61}$$

where eff-searchSP is calculated from (7.55). There is no obvious choice for the value of r. Hence, BLAST considers all possible values of r and choose the admissible set that gives the lowest P-value. This means that a set of tests are performed. For addressing this multiple testing issue, the E-value in (7.61) is further adjusted by dividing a factor of $(1 - \tau)\tau^{r-1}$. For BLASTN search, τ is set to 0.5. For other searches, τ is set to 0.5 for ungapped alignment and 0.1 for gapped alignment. Finally, the P-value for the alignments in $\mathscr{A}$ is calculated from (7.57).

Finally, we must warn that the above calculations for P-value and E-value are used in the current version of BLAST (version 2.2). They are different from what is used in the earlier versions. For example, the length adjustment was calculated as the product of λ and the raw score divided by H in the earlier version. Accordingly, they might be modified in future.

Table 7.4 The empirical values of α and β for PAM and BLOSUM matrices. Data are from Altschul et al. (2001), with permission from Oxford University Press [5].

Scoring matrix Gap cost	BLOSUM45 (14, 2)	BLOSUM62 (11, 1)	BLOSUM80 (10, 1)	PAM30 (9, 1)	PAM70 (10, 1)
α	1.92	1.90	1.07	0.48	0.70
β	-37.2	-29.7	-12.5	-5.9	8.1

7.4.2 BLAST Printouts

We now examine BLAST printouts in detail. In the case of distinct high-scoring alignments being listed, BLAST printout lists raw score and the equivalent "bit score" together with E-values calculated from (7.56). At the end, it gives the values of λ, K, and H; it also gives the length adjustment and the number of sequences and letters in database.

Let m be the length of query sequence. We use N and M to denote the number of sequences and letters in database, respectively. The current BLASTP (version 2.2.18) calculates the length adjustment $\bar{l}(s)$ for score s as an integer-valued approximation to the fixed point of the function[1]

$$g(l) = \alpha \frac{\ln(K(m-l)(M-Nl))}{\lambda} + \beta. \tag{7.62}$$

For ungapped alignment, $\alpha = \lambda/H$, and $\beta = 0$. For gapped alignment, the values of α and β depend on scoring matrix and affine gap cost. The empirical values of α and β for frequently used scoring matrices and affine gap costs are estimated in the work of Altschul et al. [5]. For convenience, these estimates are listed in Table 7.4.

The following is a partial printout from BLAST2.2.18 when yeast protein GST1 is compared against SwissProt database.

```
BLASTP 2.2.18+
...
Database: Non-redundant SwissProt sequences
332,988 sequences; 124,438,792 total letters
Query= gi|731924|sp|P40582.1|GST1_YEAST Glutathione S-transferase 1 (GST-I)
Length=234
...
Alignments
...
>sp|P46420.2|GSTF4_MAIZE Glutathione S-transferase 4 (GST-IV) (GST-27) (GST
class-phi member 4)
Length=223
```

---

[1] Empirically, the mean length $\bar{l}(s)$ of high-scoring alignment with sufficiently large score s is approximately a linear function of s (Altschul et al., 2001, [5]):

$$\bar{l}(s) = \alpha s + \beta.$$

Taking natural algorithm on both sides of (7.56),

$$s = \frac{1}{\lambda} \left[\ln\left((m-\bar{l}(s))(M-N\bar{l}(s))\right) - \ln(E) \right] \approx \frac{1}{\lambda} \ln\left((m-\bar{l}(s))(M-N\bar{l}(s))\right).$$

Hence, we have

$$\bar{l}(s) \approx \frac{\alpha}{\lambda} \ln\left((m-\bar{l}(s))(M-N\bar{l}(s))\right) + \beta.$$

...

```
   Score = 36.6 bits (83), Expect = 0.11, Method: Compositional matrix adjust.
   Identities = 29/86 (33%), Positives = 44/86 (51%), Gaps = 9/86 (10%)

Query 1    MSLPIIKVH-WLDHSRAFRLLWLLDHLNLEYEIVPYKR-DANFRAPPELKKIHPLGRSPL 58
           M+ P +KV+ W         R L  L+    ++YE+VP  R D + R P  L + +P G+ P+
Sbjct 1    MATPAVKVYGWAISPFVSRALLALEEAGVDYELVPMSRQDGDHRRPEHLAR-NPFGKVPV 59

Query 59   LEVQDRETGKKKILAESGFIFQYVLQ 84
           LE  D         L ES  I ++VL+
Sbjct 60   LEDGDL------TLFESRAIARHVLR 79


   ...


   Database: Non-redundant SwissProt sequences
        Posted date: May 23, 2008 5:56 PM
   Number of letters in database: 124,438,792
   Number of sequences in database: 332,988

Lambda      K        H
     0.320    0.137    0.401
Gapped
Lambda      K        H
     0.267    0.0410   0.140
Matrix: BLOSUM62
Gap Penalties: Existence: 11, Extension: 1
Number of Sequences: 332988
...
Length of query: 234
Length of database: 124438792
Length adjustment: 111
...
```

The printout above shows that the query sequence has length 234, the number of letters in database is 124,438,792, the number of sequences in database is 332,988, and the length adjustment is 111. The Maize Glutathione match listed in the printout contains gaps. Hence,

$$\lambda = 0.267, \ K = 0.041.$$

Because the raw is 83, the bit score is

$$[0.267 - \ln(0.041)]/\ln(2) \approx 36.58,$$

in agreement with the printout value. The value of Expect is

$$0.041 \times (234 - 111) \times (124438793 - 332988 \times 111) \times e^{-0.267 \times 83} \approx 0.105,$$

in agreement with the value 0.11 in the printout. Finally, one can easily check that the length adjustment 111 is an approximate fixed point of the function in (7.62).

Another partial printout from Wu-BLASTP search against a protein database is given below, in which only high-scoring alignments between the query and a database sequence are listed. There are eight alignments in total. For each alignment, the Karlin-Altschul sum P-value and E-value are reported. We number the alignments from 1 to 8 as they appear in the printout. It is easy to see that the alignments 1, 2, 3, 4, 6, 7 form the most significant admissible set. The P-value and E-value associated with these alignments are calculated using this set. It is not clear from the printout, however, which admissible set is used for calculating the P-values and E-values for alignments 5 and 8. Furthermore, because the values of some parameters for calculation of the E-values are missing, we are unable to verify the E-values in the printout.

```
BLASTP 2.0MP-WashU [04-May-2006] [linux26-x64-I32LPF64 2006-05-10T17:22:28]

    Copyright (C) 1996-2006 Washington University, Saint Louis, Missouri USA.
    All Rights Reserved.

    Reference: Gish, W. (1996-2006) http://blast.wustl.edu

    Query= Sequence
    (756 letters)

      ...

>UNIPROT:Q4QAI9_LEIMA Q4QAI9 Mismatch repair protein, putative. Length = 1370

    Score = 360 (131.8 bits), Expect = 5.0e-65, Sum P(6) = 5.0e-65
    Identities = 64/113 (56%), Positives = 95/113 (84%)

Query:   6  GVIRRLDETVVNRIAAGEVIQRPANAIKEMIENCLDAKSTSIQVIVKEGGLKLIQIQDNG 65
            G I +L + V+NRIAAGEV+QRP+ A+KE++EN +DA  + +QV+  EGGL+++Q+ D+G
Sbjct:   2  GSIHKLTDDVINRIAAGEVVQRPSAALKELLENAIDAGCSRVQVVAAEGGLEVLQVCDDG 61

Query:  66  TGIRKEDLDIVCERFTTSKLQSFEDLASISTYGFRGEALASISHVAHVTITTK 118
            +GI KEDL ++CER+ TSKLQ+FEDL  ++++GFRGEALASIS+V+ +T+TT+
Sbjct:  62  SGIHKEDLPLLCERYATSKLQTFEDLHRVTSFGFRGEALASISYVSRMTVTTR 114

    Score = 194 (73.4 bits), Expect = 5.0e-65, Sum P(6) = 5.0e-65
    Identities = 43/102 (42%), Positives = 59/102 (57%)

Query: 240  FKMNGYISNANYSVKKCIFLLFINHRLVESTSLRKAIETVYAAYLPKNTHPFLYLSLEIS 299
            F + GY S+  + +K   +FIN RLVES ++RKAI+ VY+  L   PF  L L +
Sbjct: 382  FTLVGYTSDPTLAQRKPYLCVFINQRLVESAAIRKAIDAVYSGVLTGGHRPFTVLLLSVP 441

Query: 300  PQNVDVNVHPTKHEVHFLHEESILERVQQHIESKLLGSNSSR 341
                VDVNVHPTK EV  L EE I+ RV +     +L + ++R
Sbjct: 442  TDRVDVNVHPTKKEVCLLDEELIVSRVAEVCRGAVLEAAAAR 483

    Score = 175 (66.7 bits), Expect = 5.0e-65, Sum P(6) = 5.0e-65
    Identities = 35/75 (46%), Positives = 49/75 (65%)

Query: 119  TADGKCAYRASYSDGKLKAPPKPCAGNQGTQITVEDLFYNIATRRKALKNPSEEYGKILE 178
            TA   A+R  Y +G L   P+PCAGN GT + VE LFYN   RR++L+  SEE+G+I++
Sbjct: 136  TAGAAVAWRCQYLNGTLLEDPQPCAGNPGTTVRVEKLFYNALVRRRSLR-ASEEWGRIVD 194

Query: 179  VVGRYSVHNAGISFS 193
            VV RY++    I F+
Sbjct: 195  VVSRYALAFPAIGFT 209

    Score = 82 (33.9 bits), Expect = 5.0e-65, Sum P(6) = 5.0e-65
    Identities = 17/66 (25%), Positives = 36/66 (54%)

Query: 603  PKEGLAEYIVEFLKKKAE---MLADYFSLEIDEEGNLIGLPLLIDNYVPP-LEGLPIFIL 658
            P++   Y+  +++      +L +YF ++I  +G L+GLP  ++ + PP +  +P+ +
Sbjct: 1125 PEDATMRYVRRLVRRLCRWRGLLKEYFYIDITADGLLVGLPYGLNRHWPPRMRAVPVMVW 1184

Query: 659  RLATEV 664
            LA  V
Sbjct: 1185 LLAEAV 1190

    Score = 72 (30.4 bits), Expect = 5.4e-64, Sum P(6) = 5.4e-64
```

```
Identities = 33/128 (25%), Positives = 53/128 (41%)

Query:  503  LTSVLSLQEEINEQGHEVLREMLHNHSFVGCVNPQWALAQHQTKLYLLNTTKLSEELFYQ 562
             L+SV + + + +      ++   SFVG V+ +  LAQ  T L   +T  L+ + +Q
Sbjct:  933  LSSVSMIVDRLLAEASPTADNLVDQLSFVGTVDSRAFLAQAGTTLLWCDTMALTRHVVFQ 992

Query:  563  ILIYDFANFG-----VLRLSEPAPLFDLAMLAL--DSPESGWTEEDGPKEGLAEYIVEFL 615
             + +          VL + P L DL +LAL  D P        P   L   + E
Sbjct:  993  RIFLRWCQPALPAPPVLAFATPVRLADLLLLALAYDGPHL-----QPPSATLLAVVEECA 1047

Query:  616  KKKAEMLA 623
             KK+ +  A
Sbjct: 1048  KKRQQQQA 1055

 Score = 70 (29.7 bits), Expect = 5.0e-65, Sum P(6) = 5.0e-65
 Identities = 17/43 (39%), Positives = 23/43 (53%)

Query:  716  VEHIVYKALRSHILP--PKHFTEDGNILQLANLPDLYKVFERC 756
             V H ++ L++  L   P      DG I  L ++ LYKVFERC
Sbjct: 1328  VRHGLFACLKNPQLCRLPDQCLRDGTIQSLVSVESLYKVFERC 1370

 Score = 56 (24.8 bits), Expect = 5.0e-65, Sum P(6) = 5.0e-65
 Identities = 40/149 (26%), Positives = 60/149 (40%)

Query:  320  ESILERV-QQHIESKLLGS--NSSRMYFTQTLLPGLAGPSGE----MVKXXXXXXXXXXX 372
             + ILE++ +QH       L S  SS +  T  + P  AG  G      +V
Sbjct:  525  QHILEKLREQHQRGAPLASPLTSSSLTSTAAVAPAGAGVGGVGPNVVVAPCTMVRVEPQK 584

Query:  373  XXXDKVYAHQMVRTDSREQKLDAFLQPLSKP--LSSQ--PQAIVTEDKTDISSGRARQQD 428
             DK Y Q+ +      A L P S P  LSS    Q I++ D      + R +
Sbjct:  585  GALDK-YFSQRLAAAAAPAAATA-LAPTSSPSSLSSSRTAQEILSRDSVP---DQLRAEA 639

Query:  429  EEMLELPAPAEVAAKNQSLEGDTTKGTSE 457
             EE L+     + +A ++ +GD T G S+
Sbjct:  640  EEPLKDGDRRQESAIQRAKKGDATNGQSQ 668

 Score = 50 (22.7 bits), Expect = 2.1e-64, Sum P(6) = 2.1e-64
 Identities = 20/88 (22%), Positives = 39/88 (44%)

Query:  383  MVRTDSREQKLDA-FLQPLSKPLS-SQPQAIV-TEDKTDISSGRARQQDEEMLELPAPAE 439
             MVR + ++  LD  F Q L+  + + +   A+  T   + +SS R  Q+      +P
Sbjct:  577  MVRVEPQKGALDKYFSQRLAAAAAPAAATALAPTSSPSSLSSSRTAQEILSRDSVPDQLR 636

Query:  440  VAAKNQSLEGDTTKGTSEMSEKRGPTSS 467
             A+   +GD + ++   K+G ++
Sbjct:  637  AEAEEPLKDGDRRQESAIQRAKKGDATN 664
 ...
```

```
Query                   ----- As Used -----   ----- Computed ----
Frame  MatID  Matrix name   Lambda   K      H      Lambda   K      H
+0     0      BLOSUM62      0.316    0.132  0.378  same     same   same
              Q=9,R=2       0.244    0.0300 0.180  n/a      n/a    n/a

Query
Frame  MatID  Length  Eff.Length    E  S W T X  E2   S2
+0     0      756     730        8.0 94 3 11 22 0.42 34
                                          37 0.45 37
```

```
Statistics:
  Database: /ebi/services/idata/v2187/blastdb/uniprot
  Title: uniprot
  Posted: 11:40:32 PM BST May 19, 2008
  Created: 11:40:32 PM BST May 19, 2008
  Format: XDF-1
  # of letters in database: 2,030,407,280
  # of sequences in database: 6,225,408
  # of database sequences satisfying E: 1265
  ...
```

7.5 Bibliographic Notes and Further Reading

There is a large body of literature on the topics presented in this chapter. We will not attempt to cite all of the literature here. Rather, we will just point out some of the most relevant and useful references on this subject matter. For further information, we refer the reader to the survey papers of Altschul et al. [4], Karlin [99], Mitrophanov and Borodovsky [141], Pearson and Wood [162], and Vingron and Waterman [193]. The book by Ewens and Grant [64] is another useful source for this subject.

7.1

Several studies have demonstrated that local similarity measures with or without gaps follow an extreme value type-I distribution. In the late 1980s and early 1990s, Arratia and Waterman studied the distribution of the longest run of matching letters. Consider two random sequences of lengths n_1 and n_2 in which the probability that two letters match is p. By generalizing the result of Erdös and RéNY [170] on the longest head run in the n tosses of a coin, Arratia and Waterman showed that the length of the longest run of matching letters in two sequences is approximately $k\log_{1/p}(n_1n_2)$ as m and n go to infinity at similar rates, where k is a constant depending on both $\ln(n_1)/\ln(n_2)$ and the letter frequencies of the sequences being compared [12, 13]. A simlar result was also proved by Karlin and Ost [103]. Later, Arratia, Gordon, and Waterman further showed that an extreme value type-I distribution even holds asymptotically for the longest run of matching letters between two sequences, allowing m mismatches [10]. In this case, the extreme value type-I distribution has the scale parameter $\log(e)$ and the location parameter

$$\log((1-p)n_1n_2) + m\log\log((1-p)n_1n_2) + m\log((1-p)/p) - \log(m!) - 1/2,$$

where $\log$ denotes logarithm base $1/p$. A similar result is also proved for the longest run of matching letters with a given proportion of mismatches [14, 11]. All these resluts can be generalized to the case of comparing multiple random sequences. The intuitive argument for the case that matches score 1 and mismatches and indels score $-\infty$ first appeared in the book by Waterman [197] (see also the book [58]).

Altschul, Dembo, and Karlin studied this problem in the more general case of aligning locally sequences with a substitution matrix [100, 102, 57]. Karlin and Dembo proved that the best segment scores follow asymptotically an extreme value type-I distribution [102]. This result is presented in Section 7.2.1. Dembo, Karlin, and Zeitouni further showed that optimal local alignment scores without gaps approach in the asymptotic limit an extreme value type-I distribution with parameters given in (7.5) and (7.6) when the condition (7.4) holds and the lengths of the sequences being aligned grow at similar rates [57].

For the coverage of the statistics of extreme values, the reader is referred to the book [50] written by Coles.

7.2

The theorems and their proofs in Sections 7.2.1 and 7.2.2 are from the work of Karlin and Dembo [102]. The Karlin-Altschul sum statistic in Section 7.2.4 is reported in [101]. The results summarized in Section 7.2.5 are found in the work of Dembo, Karlin, and Zeitouni [57]. The edge effect correction presented in Section 7.2.6 is first used in the BLAST program [7]. The justification of the edge effect correction is demonstrated by Altschul and Gish [6]. Edge effects are even more serious for gapped alignment as shown by Park and Spouge [157] and Spang and Vingron [182].

7.3

Phase transition phenomena for local similarity measures is studied by Waterman and Arratia [199, 15], Dembo, Karlin, and Zeitouni [56], and Grossmann and Yakir [82]. In general, it seems hard to characterize the logarithmic region. Toward this problem, a sufficient condition is given by Chan [39].

Many empirical studies strongly suggest that the optimal local alignment scores with gaps also follow an extreme value type-I distribution in the logarithmic zone [5, 6, 51, 144, 145, 160, 181, 200]. Although this observation seems a long way from being mathematically proved, it is further confirmed in the theoretical work of Siegmund and Yakir [179] and Grossman and Yakir [82].

Two types of methods for parameter estimation are presented in Section 7.3.2. The direct method is from the work of Altschul and Gish [6]. The island method, rooting in the work of Waterman and Vingron [200], is developed in the paper of Altschul et al. [5]. More heuristic methods for estimating the distribution parameters are reported in the papers of Bailey and Gribskov [21], Bundschuh [35], Fayyaz et al. [66], Kschischo, Lässig, and Yu [117], Metzler [137], Mott [145], and Mott and Tribe [146].

7.4

The material covered in Section 7.4.1 can be found in the manuscript of Gertz [74]. The printouts given in Section 7.4.1 are prepared using NCBI BLAST and EBI WU-BLAST web server, respectively.

Miscellaneous

We have studied the statistical significance of local alignment scores. The distribution of global alignment scores is hardly studied. No theoretical result is known. Readers are referred to the papers by Reich et al. [169] and Webber and Barton [201] for information on global alignment statistics.

Chapter 8
Scoring Matrices

With the introduction of the dynamic programming algorithm for comparing protein sequences in 1970s, a need arose for scoring amino acid substitutions. Since then, the construction of scoring matrices has become one key issue in sequence comparison. A variety of considerations such as the physicochemical and three-dimensional structure properties have been used for deriving amino acid scoring (or substitution) matrices.

The chapter is divided into eight sections. The PAM matrices are introduced in Section 8.1. We first define the PAM evolutionary distance. We then describe the Dayhoff's method of constructing the PAM matrices. Frequently used PAM matrices are listed at the end of this section.

In Section 8.2, after briefly introducing the BLOCK database, we describe the Henikoff and Henikoff's method of constructing the BLOSUM matrices. In addition, frequently used BLOSUM matrices are listed.

In Section 8.3, we show that in seeking local alignment without gaps, any amino acid scoring matrix takes essentially a log-odds form. There is a one-to-one correspondence between the so-called valid scoring matrices and the sets of target and background frequencies. Moreover, given a valid scoring matrix, its implicit target and background frequencies can be retrieved efficiently.

The log-odds form of the scoring matrices suggests that the quality of database search results relies on the proper choice of scoring matrix. Section 8.4 describes how to select scoring matrix for database search with a theoretic-information method.

In comparison of protein sequences with biased amino acid compositions, standard scoring matrices are no longer optimal. Section 8.5 introduces a general procedure for converting a standard scoring matrix into one suitable for the comparison of two sequences with biased compositions.

For certain applications of DNA sequence comparison, nontrivial scoring matrix is critical. In Section 8.6, we discuss a variant of the Dayhoff's method in constructing nucleotide substitution matrices. In addition, we address why comparison of protein sequences is often more effective than that of the coding DNA sequences.

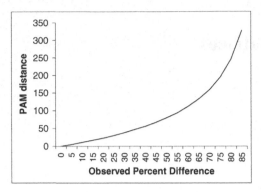

Fig. 8.1 The correspondence between the PAM evolutionary distances and the dissimilarity levels.

There is no general theory on constructing scoring matrices for local alignments with gaps. One issue in gapped alignment is to select appropriate gap costs. In Section 8.7, we briefly discuss the affine gap cost model and related issues in gapped alignment.

We conclude the chapter with the bibliographic notes in Section 8.8.

8.1 The PAM Scoring Matrices

The PAM matrices are the first amino acid substitution matrices for protein sequence comparison. Theses matrices were first constructed by Dayhoff and coworkers based on a Markov chain model of evolution. A *point accepted mutation* in a protein is a substitution of one amino acid by another that is "accepted" by natural selection. For a mutation to be accepted, the resulting amino acid must have the same function as the original one. A *PAM unit* is an evolutionary time period over which 1% of the amino acids in a sequence are expected to undergo accepted mutations. Because a mutation might occur several times at a position, two protein sequences that are 100 PAM diverged are not necessarily different in every position; instead, they are expected to be different in about 52% of positions. Similarly, two protein sequences that are 250 PAM diverged have only roughly 80% dissimilarity. The correspondence between the PAM evolutionary distance and the dissimilarity level is shown in Figure 8.1.

Dayhoff and her coworkers first constructed the amino acid substitution matrix for one PAM time unit, and then extrapolated it to other PAM distances. The construction started with 71 blocks of aligned protein sequences. In each of these blocks, a sequence is no more than 15% different from any other sequence. The high within-block similarity was imposed to minimize the number of substitutions that may have resulted from multiple substitutions at the same position.

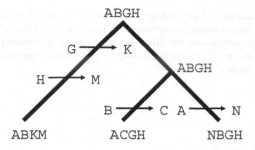

Fig. 8.2 A phylogenetic tree over three observed protein sequences. Inferred ancestors are shown at the internal nodes, and amino acid substitutions are indicated along the branches.

In the original construction of Dayhoff et al., a phylogenetic tree was constructed over the sequences in each block, and symmetric substitutions along the branches were counted by the way illustrated in Figure 8.2. In total, 1572 substitutions in Table 8.1 were observed in 71 phylogenetic trees.

Assume that the amino acids are numbered from 1 to 20 and denote the (i, j)-th entry in the substitution matrix (in Table 8.1) by A_{ij}. For $i \neq j$, we define

$$p_{ij} = c \frac{A_{ij}}{\sum_k A_{ik}},$$

where c is a positive constant to be specified according to the constraint of one PAM divergence, and set

$$p_{ii} = 1 - \sum_{j \neq i} p_{ij}.$$

Note that $\sum_j p_{ij} = 1$. If c is chosen to be small enough so that each p_{ii} is non-negative, the matrix (p_{ij}) can be considered as the transition matrix of a Markov chain.

Let f_i be the frequency of the ith amino acid in the sequences appearing in the 71 phylogenetic trees. Then, the expected proportion of amino acids that change after one PAM is

$$\sum_i f_i \sum_{j \neq i} p_{ij} = c \sum_i \sum_{j \neq i} f_i \left(A_{ij} / \sum_k A_{ik} \right).$$

Because 1% of amino acids are expected to change over the one PAM period, we set c as

$$c = \frac{0.01}{\sum_i \sum_{j \neq i} f_i \left(A_{ij} / \sum_k A_{ik} \right)}.$$

By estimating f_i with the observed frequency of the corresponding amino acid in all the sequences in the 71 phylogenetic trees, Dayhoff et al. obtained the substitution matrix M_1 over the one PAM period, which is given in Table 8.2.

If $M_1 = (m_{ij})$ is considered as the transition matrix of the Markov chain model of the evolution over one PAM period, then M_1^n is the transition matrix over the n

Table 8.1 The 1572 amino acid substitutions were observed in the 71 phylogenetic trees. Here, each entry is 10 times the corresponding substitutions. Fractional substitution numbers result when ancestral sequences are ambiguous, for which the substitutions are counted statistically. This is reproduced from the paper [55] of Dayoff, Schwartz, and Orcutt; reprinted with permission of Nat'l Biomed. Res. Foundation.

	A	R	N	D	C	Q	E	G	H	I	L	K	M	F	P	S	T	W	Y	V
Ala A																				
Arg R	30																			
Asn N	109	17																		
Asp D	154	0	532																	
Cys C	33	10	0	0																
Gln Q	93	120	50	76	0															
Glu E	266	0	94	831	0	422														
Gly G	579	10	156	162	10	30	112													
His H	21	103	226	43	10	243	23	10												
Ile I	66	30	36	13	17	8	35	0	3											
Leu L	95	17	37	0	0	75	15	17	40	253										
Lys K	57	477	322	85	0	147	104	60	23	43	39									
Met M	29	17	0	0	0	20	7	7	0	57	207	90								
Phe F	20	7	7	0	0	0	0	17	20	90	167	0	17							
Pro P	345	67	27	10	10	93	40	49	50	7	43	43	4	7						
Ser S	772	137	432	98	117	47	86	450	26	20	32	168	20	40	269					
Thr T	590	20	169	57	10	37	31	50	14	129	52	200	28	10	73	696				
Trp W	0	27	3	0	0	0	0	0	3	0	13	0	0	10	0	17	0			
Tyr Y	20	3	36	0	30	0	10	0	40	13	23	10	0	260	0	22	23	6		
Val V	365	20	13	17	33	27	37	97	30	661	303	17	77	10	50	43	186	0	17	

PAM period. Let $M_1^n = \left(m_{ij}^{(n)} \right)$. The entry $m_{ij}^{(n)}$ is the probability that the ith amino acid is replaced by the jth amino acid in a position over the n PAM period. As n gets larger, all the entries in the jth column of this matrix have approximately the same value, converging to the background frequency f_j of the jth amino acid. Finally, the (i, j)-entry in the PAMn matrix is defined to be

$$C \log \left(m_{ij}^{(n)} / f_j \right)$$

where C is set to 10 in the original construction of Dayhoff et al.

Although the PAM250 matrix shown in Figure 8.3 was the only matrix published by Dayhoff et al., other PAM matrices can be obtained in the same way. PAM30, PAM70, and PAM120 given in Figures 8.4 – 8.6 are also frequently used for homology search.

The PAM matrices have visible patterns. For instance, they show the three groups of chemically similar amino acids that tend to replace one another: the basic group - H, R, K; the aromatic group - F, Y, W; the acid and acid-amide group - D, E, N, Q; the hydrophobic group - M, I, L, V.

In literature, the PAM matrices sometimes include four extra rows/columns denoted by B, Z, X, and *. The B-entries are the frequency-weighted average of the D(Asp)- and N(Asn)-entries; the Z-entries are the frequency-weighted average of Q(Gln)- and E(Glu)-entries; the X-entries are the frequency-weighted average of all

Table 8.2 The amino acid substitution matrix over the one PAM period. The (i, j)-entry equals 10,000 times the probability that the amino acid in column j is replaced by the amino acid in row i. This is reproduced from the paper [55] of Dayoff, Schwartz, and Orcutt; reprinted with permission of Nat'l Biomed. Res. Foundation.

	A	R	N	D	C	Q	E	G	H	I	L	K	M	F	P	S	T	W	Y	V
A	9876	2	9	10	3	8	17	21	2	6	4	2	6	2	22	35	32	0	2	18
R	1	9913	1	0	1	10	0	0	10	3	1	19	4	1	4	6	1	8	0	1
N	4	1	9822	36	0	4	6	6	21	3	1	13	0	1	2	20	9	1	4	1
D	6	0	42	9859	0	6	53	6	4	1	0	3	0	0	1	5	3	0	0	1
C	1	1	0	0	9973	0	0	0	1	1	0	0	0	0	1	5	1	0	3	2
Q	3	9	4	5	0	9876	27	1	23	1	3	6	4	0	6	2	2	0	0	1
E	10	0	7	56	0	35	9865	4	2	3	1	4	1	0	3	4	2	0	1	2
G	21	1	12	11	1	3	7	9935	1	0	1	2	1	1	3	21	3	0	0	5
H	1	8	18	3	1	20	1	0	9912	0	1	1	0	2	3	1	1	1	4	1
I	2	2	3	1	2	1	2	0	0	9872	9	2	12	7	0	1	7	0	1	33
L	3	1	3	0	0	6	1	1	4	22	9947	2	45	13	3	1	3	4	2	15
K	2	37	25	6	0	12	7	2	2	4	1	9926	20	0	3	8	11	0	1	1
M	1	1	0	0	0	2	0	0	0	5	8	4	9874	1	0	1	2	0	0	4
F	1	1	1	0	0	0	0	1	2	8	6	0	4	9946	0	2	1	3	28	0
P	13	5	2	1	1	8	3	2	5	1	2	2	1	1	9926	12	4	0	0	2
S	28	11	34	7	11	4	6	16	2	2	1	7	4	3	17	9840	38	5	2	2
T	22	2	13	4	1	3	2	2	1	11	2	8	6	1	5	32	9871	0	2	9
W	0	2	0	0	0	0	0	0	0	0	0	0	0	1	0	1	0	9976	1	0
Y	1	0	3	0	3	0	1	0	4	1	1	0	0	21	0	1	1	2	9945	1
V	13	2	1	1	3	2	2	3	3	57	11	1	17	1	3	2	10	0	2	9901

the entries for the 20 amino acids; * stands for any character that is not an amino acid, such as the translation * of an end codon.

8.2 The BLOSUM Scoring Matrices

For homology search, what we actually do is to test whether two residues are correlated due to the fact that they are homologous or not. The probability that two residues are aligned in an alignment of two homologous sequences is called the target frequency. If two residues are uncorrelated, occurring independently, the probability that we expect to observe these two residues aligned is the product of the probabilities that these two residues occur in a sequence. This product is called the background frequency. As we shall see in the next section, theory says that the best score for aligning two amino acids is essentially the logarithm of the ratio of their target frequencies to their background frequencies. In 1992, Henikoff and Henikoff constructed the BLOSUM scoring matrices from the target frequencies inferred from roughly 2000 blocks. Here, each block is the ungapped alignment of a conserved region of a protein family. These blocks were obtained from 500 protein groups.

Henikoff and Henikoff first counted the number of occurrences of each amino acid and the number of occurrences of each pair of amino acids aligned in the block dataset. Assume a block has k sequences of length w. There are kw residues and $wk(k-1)/2$ pairs of aligned amino acids in this block. If amino acids i and j occur p and q times in a column, respectively, then there are $\binom{p}{2}$ i-i pairs and pq i-j pairs in the column. The counts of all the possible pairs in every column of each block in the dataset are then summed.

We further use n_{ij} to denote the number of i-j pairs in the block dataset. The frequency that we expect to see i and j aligned is

$$p_{ij} = \frac{n_{ij}}{\sum_{k \leq l} n_{lk}}.$$

The frequency p_i that i occurs somewhere in the block is calculated as

$$p_i = p_{ii} + \sum_{j < i} \frac{p_{ij}}{2}.$$

Given the frequencies p_k for all $1 \leq k \leq 20$, the probability e_{ij} that two amino acids i and j are aligned by chance is $p_i p_j$ if $i = j$ and $2 p_i p_j$ otherwise. A BLOSUM matrix is obtained by taking 2 times the logarithm of p_{ij}/e_{ij} to base 2 and rounding to the nearest integer.

The score for each pair of amino acids in a BLOSUM matrix is positive if they are more likely than chance, and negative if they are less likely.

This counting approach overlooks an important factor that can bias the result. If there are many very closely related proteins and a few others that are less closely related in a block, then the contribution of that block will be biased toward the closely related proteins. As a result, the substitution matrix derived will not be good for detecting two distantly related protein sequences. To reduce multiple contributions to the frequencies of pairs of amino acids, sequences in a block are first clustered. This is done by specifying a cutoff similarity level $x\%$ and then grouping the sequences in each block into clusters in such a way that each sequence in a cluster has $x\%$ or higher similarity to one or more other sequences in the same cluster.

The sequences in each cluster are then weighted as a single sequence in frequency calculation. Specifically, each occurrence of an amino acid in a sequence is counted as $\frac{1}{m}$ times, where m is the size of the cluster that contains the sequence. Each occurrence of an amino acid pair within a cluster is not counted; it is counted as $\frac{1}{mn}$ times if they are from different clusters, where m and n are the sizes of the two clusters from which the two sequences are taken.

If $x\%$ is used as the cutoff similarity level for clustering, the resulting substitution matrix is called the BLOSUMx matrix. BLOSUM62 given in Figure 8.7 is the most frequently used scoring matrix for homology search. Other popular scoring matrices include BLOSUM45 and BLOSUM80, which are given in Figures 8.8 and 8.9. The BLOSUM matrices seen in literature sometimes include four extra rows/columns denoted by B, Z, X, and *. These four columns have the same meaning as in the PAM matrices.

The program used for constructing the block dataset uses scoring matrix, too! This raises two questions: what substitution matrices are used there and whether those matrices bias the result or not? To break the circularity and eliminate the bias effect, Henikoff and Henikoff took a three-step iterative approach. First, a unitary scoring matrix, where the match score is 1 and the mismatch score is 0, was used

initially, generating 2205 blocks; a scoring matrix was obtained from these blocks
by clustering at similarity level 60%. Next, the resulting scoring matrix was used to
construct a second dataset of 1961 blocks, and another scoring matrix was obtained
in the same way as in the first step. Finally, the second scoring matrix was used to
construct the final version of the dataset of 2106 blocks (the BLOCKS database,
version 5.0); from this final dataset, various BLOSUM matrices were constructed
using corresponding similarity levels.

8.3 General Form of the Scoring Matrices

We now consider optimal ungapped local alignments obtained with scoring matrix
(s_{ij}) for the equences in a protein model that is specified by a set of background
frequencies p_i for all amino acids i. Assume (s_{ij}) has at least one positive entry and
the expected score $\sum_{ij} p_i p_j s_{ij}$ for aligning two random residues is negative. Karlin
and Altschul (1990, [100]) showed that among optimal ungapped local alignments
obtained from comparison of random sequences with scoring matrix (s_{ij}), the amino
acids a_i and a_j are aligned with frequency

$$q_{ij} = p_i p_j e^{\lambda s_{ij}} \tag{8.1}$$

approximately. Rewriting (8.1), we have

$$s_{ij} \approx \frac{1}{\lambda} \ln\left(\frac{q_{ij}}{p_i p_j}\right). \tag{8.2}$$

Therefore, the scoring matrix (s_{ij}) has an implicit set of target frequencies q_{ij} for
aligning amino acids satisfying (8.2). In other words, the substitution score for a pair
of amino acids in any scoring matrix is essentially a log-odds ratio. Accordingly, no
matter what method is used, the resulting scoring matrices implicitly have the same
underlying mathematical structure as the Dayhoff and BLOSUM matrices.

Consider a scoring matrix (s_{ij}) and a protein model given by background frequencies p_i's. As long as the expected score $\sum_{ij} p_i p_j s_{ij}$ remains negative, (s_{ij}) can
always be expressed in the log-odds form

$$s_{ij} = \frac{1}{\lambda} \ln\left(\frac{z_{ij}}{p_i p_j}\right)$$

for some z_{ij} ($1 \leq i, j \leq 20$). However, the equations

$$p_i = \sum_j z_{ij}$$

and

$$p_j = \sum_i z_{ij}$$

may not necessarily hold.

A scoring matrix (s_{ij}) is *valid* if there is a set of target frequencies q_{ij}, summing to 1, such that

$$s_{ij} = \frac{1}{\lambda} \ln \left(\frac{q_{ij}}{p_i p'_j} \right) \tag{8.3}$$

for some λ, where p_i and p'_j are the marginal sums of the q_{ij}:

$$p_i = \sum_j q_{ij}, \tag{8.4}$$

$$p'_j = \sum_i q_{ij}. \tag{8.5}$$

In this definition, the matrix (q_{ij}) is not requested to be symmetric.

It can be proved that a valid scoring matrix corresponds uniquely to a set of target frequencies with their implied background frequencies (Yu, Wootton, and Altschul, 2003, [211]). In what follows, we only show how to find the associated set of target frequencies from a valid matrix.

We rewrite (8.3) as

$$p_i e^{\lambda s_{ij}} = \frac{q_{ij}}{p'_j}.$$

By (8.5),

$$\sum_i p_i e^{\lambda s_{ij}} = 1, \ \ 1 \le j \le 20. \tag{8.6}$$

Define $M(\lambda) = (e^{\lambda s_{ij}})$. $M(\lambda)$ is a 20×20 functional matrix with variable λ and (8.6) becomes

$$M(\lambda)^T (p_1, p_2, \cdots, p_{20})^T = (1, 1, \cdots, 1)^T.$$

Let the formal inverse of $M(\lambda)$ be $Y(\lambda) = (Y_{ij}(\lambda))$. We then have

$$p_i = \sum_j Y_{ji}(\lambda), \ \ 1 \le i \le 20. \tag{8.7}$$

Similarly,

$$p'_j = \sum_i Y_{ji}(\lambda), \ \ 1 \le j \le 20. \tag{8.8}$$

Finally, the condition $\sum_i p_i = 1$ implies that

$$\sum_{i,j} Y_{ij}(\lambda) = 1. \tag{8.9}$$

Solving (8.9), we obtain the value of λ. Once λ is known, we obtain p_i, p'_j, and q_{ij} from (8.7), (8.8), and (8.3) for all possible i and j, respectively.

8.4 How to Select a Scoring Matrix?

From the discussion in Section 8.3, given a protein sequence model in which amino acids occur by chance with background frequencies p_i $(1 \le i \le 20)$, a scoring matrix s_{ij} with negative expected score and at least one positive entry is only optimal to the alignments in which the amino acids a_i and a_j are aligned with target frequencies q_{ij} that satisfy equation (8.2). In other words, different scoring matrices are optimal for detecting different classes of alignments. This raises the problem of selecting scoring matrix for best distinguishing true protein alignment from chance.

Multiplying a scoring matrix by a positive constant has no effect on the relative scores of different MSPs. Therefore, two matrices related by such a constant factor are said to be equivalent. By (8.2), any scaling corresponds merely to taking logarithm to a different base. When λ is set to 1, the scores are natural logarithms of the odds ratios; when λ is set to $\ln 2$, the scores become logarithms to base 2.

Let N denote the expected number of MSPs with score s or more obtained in the alignment of two sequences of length n_1 and n_2. Setting λ to $\ln 2$ in (7.45), we obtain

$$s = \log_2 \left(\frac{K}{N} \right) + \log_2(n_1 n_2). \qquad (8.10)$$

The parameter K is usually less than 1 for a typical scoring matrix. An alignment may be considered significant when N is 0.01. As a result, the right-hand side of equation (8.10) is dominated by the term $\log_2(mn)$, and alignment score needed to distinguish a MSP from chance is approximately $\log_2(mn)$. Thus, for comparing two proteins of 250 amino acid residues, a MSP is statistically significant only if its score is 16 ($\approx 2\log_2 250$) or more; if such a protein is searched against a database of 10,000,000 residues, a significant MSP should have then score 31 ($\approx \log_2 2500000000$) or more.

The PAM and BLOSUM matrices given in Sections 8.1 and 8.2 have different λs (see Table 8.3). By multiplying $\lambda / \ln 2$, a scoring matrix is normalized into the logarithms of odds ratios to base 2. Scores obtained with such a normalized scoring matrix can be considered as bit information. As a result, we may say that 16-bit information is required to distinguish a MSP from chance in comparing two protein sequences of 250 amino acid residues.

Given a protein model (which specifies the background frequencies p_i) and a normalized scoring matrix (s_{ij}), one can calculate the target frequencies q_{ij}, which characterize the alignments on which the scoring matrix is optimal, as

$$q_{ij} = p_i p_j e^{(\ln 2)s_{ij}}.$$

Define

$$H = \sum_{i,j} q_{ij} s_{ij} = \sum_{i,j} q_{ij} \log_2 \left(\frac{q_{ij}}{p_i p_j} \right). \qquad (8.11)$$

H is the expected score (or bit information) per residue pair in the alignments characterized by the target frequencies. From theoretic-information point of view, H is

Table 8.3 The values of the parameter λ in equation (8.2) and the relative entropy of PAM and BLOSUM matrices listed in Sections 8.1 and 8.2.

	PAM30	PAM70	PAM120	PAM250	BLOSUM45	BLOSUM62	BLOSUM80
λ	$\ln 2/2$	$\ln 2/2$	$\ln 2/2$	$\ln 10/10$	$\ln 10/10$	$\ln 2/2$	$\ln 10/10$
Entropy	2.57	1.60	0.979	0.354	0.3795	0.6979	0.9868

the relative entropy of the target frequency distribution with respect to the background distribution (see Section B.6). Hence, we call H the relative entropy of the scoring matrix (s_{ij}) (with respect to the protein model). Table 8.3 gives the relative entropy of popular scoring matrices with respect to the implicit protein model.

Intuitively, if the value of H is high, relatively short alignments with the target frequencies can be distinguished from chance; if the value of H is low, however, long alignments are necessary. Recall that distinguishing an alignment from chance needs 16 bits of information in comparison of two protein sequences of 250 amino acids. Using this fact, we are able to estimate the length of a significant alignment of two sequences that are x-PAM divergent. For example, at a distance of 120 PAMs, there is on average 0.979 bit of information in every aligned position as shown in Table 8.3. As a result, a significant alignment has at least 17 residues.

For database search, the situation is more complex. In this case, alignments are unknown and hence it is not clear which scoring matrix is optimal. It is suggested to use multiple PAM matrices.

The PAMx matrix is designed to compare two protein sequences that are separated by x PAM distance. If it is used to compare two protein sequences that are actually separated by y PAM distance, the average bit information achieved per position is smaller than its relative entropy. When y is close to x, the average information achieved is near-optimal. Assume we are satisfied with using a PAM matrix that yields a score greater than 93% of the optimal achievable score. Because a significant MSP contains about 31 bits of information in searching a protein against a protein database containing 10,000,000 residues, the length range of the local alignments that the PAM120 matrix can detect is from 19 to 50. As a result, when PAM120 is used, it may miss short but strong or long but weak alignments that contain sufficient information to be found. Accordingly, PAM40 and PAM250 may be used together with PAM120 in database search.

8.5 Compositional Adjustment of Scoring Matrices

We have showed that a scoring matrix is only valid in one unique context in Section 8.3. Thus, it is not ideal to use a scoring matrix that is constructed for a specific set of target and background frequencies in a different context. To compare proteins having biased composition, one approach is to repeat Henikoff and Henikoff's procedure of construction of the BLOSUM matrices. A set of true alignments for the proteins under consideration is first constructed. From this alignment set, a new scoring matrix is then derived. But there are two problems with this approach. First, it requires a large set of alignments. Such a set is often not available. Second, the whole procedure requires a curatorial effort. Accordingly, an automatic adjustment of a standard scoring matrix for different compositions is necessary. In the rest of this section, we present a solution to this adjustment problem, which is due to Yu and Altschul (2005, [212]).

Consider a scoring matrix (s_{ij}) with implicit target frequencies (q_{ij}) and a set of background frequencies (P_i) and (P'_j) that are inconsistent with (q_{ij}). Here, (P_i) and (P'_j) are not necessarily equal although they are in the practical cases of interest. The problem of adjusting a scoring matrix is formulated to find a set of target frequencies (Q_{ij}) that minimize the following relative entropy with respect to the distribution (q_{ij})

$$D((Q_{ij})) = \sum_{ij} Q_{ij} \ln \left(\frac{Q_{ij}}{q_{ij}} \right) \qquad (8.12)$$

subject to consistency with the given background frequencies (P_i) and $\left(P'_j \right)$:

$$\sum_j Q_{ij} = P_i, \ 1 \le i \le 20 \qquad (8.13)$$

$$\sum_i Q_{ij} = P'_j, \ 1 \le j \le 20 \qquad (8.14)$$

Because the Q_{ij}, P_i, and P_j sum to 1 respectively, (8.13) and (8.14) impose 39 independent linear constraints on the Q_{ij}. Because

$$\frac{\partial^2 D}{\partial^2 Q_{ij}} = \frac{1}{Q_{ij}} > 0,$$

and

$$\frac{\partial^2 D}{\partial Q_{ij} \partial Q_{km}} = 0, \ i \ne k \text{ or } j \ne m,$$

the problem has a unique solution under the constraints of (8.13) and (8.14).

An additional constraint to impose is to keep the relative entropy H of the scoring matrix sought unchanged in the given background:

$$\sum_{ij} Q_{ij} \ln \left(\frac{Q_{ij}}{P_i P'_j} \right) = \sum_{ij} q_{ij} \ln \left(\frac{q_{ij}}{P_i P'_j} \right). \qquad (8.15)$$

Now we have a non-linear optimization problem. To find its optimal solution, one may use Lagrange multipliers. In non-linear optimization theory, the method of Lagrange multipliers is used for finding the extrema of a function of multiple variables subject to one or more constraints. It reduces an optimization problem in

k variables with m constraints to a problem in $k+m$ variable with no constraints. The new objective function is a linear combination of the original objective function and the m constraints in which the coefficient of each constraint is a scalar variable called the Lagrange multiplier.

Here we introduce 20 Lagrange multipliers α_i for the constraints in (8.13), 19 Lagrange multipliers β_j for the first 19 constraints in (8.14), and additional Lagrange multiplier γ for the constraint in (8.15). To simplify our description, we define $\beta_{20} = 0$. Consider the Lagrangian

$$F\left((Q_{ij}),(\alpha_i),(\beta_j),\gamma\right)$$
$$= D(Q_{ij}) + \sum_i \alpha_i \left(P_i - \sum_j Q_{ij}\right) + \sum_j \beta_j \left(P'_j - \sum_i Q_{ij}\right)$$
$$+\gamma\left[\sum_{ij} q_{ij}\ln\left(\frac{q_{ij}}{P_iP'_j}\right) - \sum_{ij} Q_{ij}\ln\left(\frac{Q_{ij}}{P_iP'_j}\right)\right]. \tag{8.16}$$

Setting the partial derivative of the Lagrangian F with respect to each of the Q_{ij} equal to 0, we obtain that

$$\ln\left(\frac{Q_{ij}}{q_{ij}}\right) + 1 - \left(\alpha_i + \beta_j + \gamma\left(\ln\left(\frac{Q_{ij}}{P_iP'_j}\right)+1\right)\right) = 0. \tag{8.17}$$

The multidimensional Newtonian method may be applied to equations (8.13) – (8.15) and (8.17) to obtain the unique optimal solution (Q_{ij}).

After (Q_{ij}) is found, we calculate the associated scoring matrix (S_{ij}) as

$$S_{ij} = \frac{1}{\lambda}\ln\left(\frac{Q_{ij}}{P_iP'_j}\right),$$

which has the same λ as the original scoring matrix (s_{ij}). The constraint (8.17) may be rewritten as

$$Q_{ij} = e^{(\alpha_i-1)/(1-\gamma)}e^{(\beta_j+\gamma)/(1-\gamma)}q_{ij}^{1/(1-\gamma)}\left(P_iP'_j\right)^{-\gamma/(1-\gamma)}.$$

Table 8.4 PAM substitution scores (bits) calculated from (8.18) in the uniform model.

PAM distance	Match score	Mismatch score	Information per position
5	1.928	-3.946	1.64
30	1.588	-1.593	0.80
47	1.376	-1.096	0.51
70	1.119	-0.715	0.28
120	0.677	-0.322	0.08

This implies that the resulting scoring matrix (S_{ij}) related to the original scores (s_{ij}) by

$$S_{ij} = \gamma' s_{ij} + \delta_i + \varepsilon_j,$$

for some γ', δ_i, and ε_j. Here, we omit the formal argument of this conclusion. For details, the reader is referred to the work of Yu and Altschul (2005, [212]).

8.6 DNA Scoring Matrices

We have discussed how scoring matrices are calculated in the context of protein sequence comparison. All the results carry through to the case of DNA sequence comparison. In particular, one can construct proper scoring matrix for aligning DNA sequences in an non-standard context using the log-odds approach as follows.

In comparison of non-coding DNA sequences, two evolutionary models are frequently used. One assumes all nucleotides are uniformly distributed and all substitutions are equally likely. This model yields a scoring scheme by which all the matches have the same score and so do all the mismatches. The so-called transition-transversion model assumes that transitions (A $\leftrightarrow$ G and C $\leftrightarrow$ T) are threefold more likely than transversions (A $\leftrightarrow$ C, A $\leftrightarrow$ T, G $\leftrightarrow$ C, and G $\leftrightarrow$ T). As a result, transitions and transversions score differently.

Let M be the transition probability matrix that reflects 99% sequence conservation and one point accepted mutation per 100 bases (1 PAM distance) and let $\alpha = 0.01$. In the uniform distribution model,

$$M = \begin{pmatrix} 1-\alpha & \frac{1}{3}\alpha & \frac{1}{3}\alpha & \frac{1}{3}\alpha \\ \frac{1}{3}\alpha & 1-\alpha & \frac{1}{3}\alpha & \frac{1}{3}\alpha \\ \frac{1}{3}\alpha & \frac{1}{3}\alpha & 1-\alpha & \frac{1}{3}\alpha \\ \frac{1}{3}\alpha & \frac{1}{3}\alpha & \frac{1}{3}\alpha & 1-\alpha \end{pmatrix}.$$

In the transition-transversion model, a transition is threefold more likely than a transversion and hence

$$M = \begin{pmatrix} 1-\alpha & \frac{3}{5}\alpha & \frac{1}{5}\alpha & \frac{1}{5}\alpha \\ \frac{3}{5}\alpha & 1-\alpha & \frac{1}{5}\alpha & \frac{1}{5}\alpha \\ \frac{1}{5}\alpha & \frac{1}{5}\alpha & 1-\alpha & \frac{3}{5}\alpha \\ \frac{1}{5}\alpha & \frac{1}{5}\alpha & \frac{3}{5}\alpha & 1-\alpha \end{pmatrix},$$

where the off-diagonal elements corresponding to transitions are $\frac{3}{5}\alpha$ and those for transversions are $\frac{1}{5}\alpha$.

In a Markov chain evolutionary model, the matrix of probabilities for substituting base i by base j after n PAMs is calculated by n successive iterations of M:

$$(m_{ij}) = (M)^n.$$

The n-PAM score s_{ij} for aligning base i with base j is simply the logarithm of the relative chance of the pair occurring in an alignment of homologous sequences as opposed to that occurring in a random alignment with background frequencies:

$$s_{ij} = \log \left(\frac{p_i m_{ij}}{p_i p_j} \right) = \log (4m_{ij}) \qquad (8.18)$$

because both models assume equal frequencies for the four bases.

Table 8.4 shows substitution scores for various PAM distances in the uniform model. Base 2 is used for these calculations, and hence these scores can be thought of as bit information. At 47 PAMs (about 65% sequence conservation), the ratio of the match and mismatch scores is approximately 5 to 4; the resulting scores are equivalent to those used in the current version of the BLASTN.

Table 8.5 presents the substitution scores for various PAM distances in the transition-transversion model. Notice that for 120 PAM distance, transitions score positively and hence are considered as conservation substitutions. Numerical calculation indicates that transitions score positively for 87 PAMs or more.

All derived PAM substitution scores can be used to compare non-coding DNA sequences. The best scores to use will depend upon whether one is seeking distantly or closely related sequences. For example, in sequence assembly, one often uses alignment to determine whether a new segment of sequence overlaps an existing sequence significantly or not to form a contig. In this case, one is interested only in alignments that differ by a few bases. Hence, PAM-5 substitution scores are more effective than other scores. On the other hand, PAM120 scoring matrix is probably more suitable in the application of finding transcriptor-binding sites.

One natural question often asked by a practitioner of homology search is: Should I compare gene sequences or the corresponding protein sequence? This can be answered through a simple quantitative analysis. Synonymous mutations are nucleotide substitutions that do not result in a change to the amino acids sequence of a protein. Evolutionary study suggests that there tend to be approximately 1.5 synonymous point mutations for every nonsynonymous point mutation. Because each codon has 3 nucleotides, each protein PAM translates into roughly $\frac{1+1.5}{3} \approx 0.8$ PAMs in DNA level. In the alignment of two proteins that have diverged by 120 protein PAMs, each residue carries on average 0.98-bit information (see Ta-

Table 8.5 PAM substitution scores (bits) calculated from (8.18) in the transition-transversion model.

PAM distance	Match score	Transition score	Transversion score	Information per position
5	1.928	-3.113	-4.667	1.65
30	1.594	-0.854	-2.223	0.85
47	1.391	-0.418	-1.669	0.57
70	1.148	-0.115	-1.217	0.35
120	0.740	0.128	-0.693	0.13

ble 8.3), whereas in the alignment of two DNA sequences that are diverged at 96 (or 120×0.8) PAMs, every three residues (a codon) carry only about 0.62-bit information. In other words, at this evolutionary distance, as much as 37% of the information available in protein comparison will be lost in DNA sequence comparison.

8.7 Gap Cost in Gapped Alignments

There is no general theory available for guiding the choice of gap costs. The most straightforward scheme is to charge a fixed penalty for each indel. Over the years, it has been observed that the optimal alignments produced by this scheme usually contain a large number of short gaps and are often not biologically meaningful (see Section 1.3 for the definition of gaps).

To capture the idea that a single mutational event might insert or delete a sequence of residues, Waterman and Smith (1981, [180]) introduced the affine gap penalty model. Under this model, the penalty $o + e \times k$ is charged for a gap of length k, where o is a large penalty for opening a gap and e a smaller penalty for extending it. The current version of BLASTP uses, by default, 11 for gap opening and 1 for gap extension, together with BLOSUM62, for aligning protein sequences.

The affine gap cost is based on the hypothesis that gap length has an exponential distribution, that is, the probability of a gap of length k is $\alpha(1-\beta)\beta^k$ for some constant α and β. Under this hypothesis, an affine gap cost is derived by charging $\log(\alpha(1-\beta)\beta^k)$ for a gap of length k. But, this hypothesis might not be true in general. For instance, the study of Benner, Cohen, and Gonnet (1993, [26]) suggests that the frequency of a length k is accurately described by $mk^{-1.7}$ for some constant m.

A generalized affine gap cost is introduced by Altschul (1998, [2]). A *generalized gap* consists of a consecutive sequence of indels in which spaces can be in either row. A generalized gap of length 10 may contain 10 insertions; it may also contain 4 insertions and 6 deletions. To reflect the structural property of a generalized gap, a generalized affine gap cost has three parameters a, b, c. The score $-a$ is introduced for the opening of a gap; $-b$ is for each residue inserted or deleted; and $-c$ is for each pair of residues left unaligned. A generalized gap with k insertions and l deletions scores $-(a + |k-l|b + c\min\{k,l\})$.

Generalized affine gap costs can be used for aligning locally or globally protein sequences. The dynamic programming algorithm carries over to this generalized affine gap cost in a straightforward manner; and it still has quadratic time complexity. For local alignment, the distribution of optimal alignment scores also follows approximately an extreme value distribution (7.1). The empirical study of Zachariah et al. (2005, [213]) shows that this generalized affine gap cost model improves significantly the accuracy of protein alignment.

8.8 Bibliographic Notes and Further Reading

For DNA sequence comparison and database search, simple scoring schemes are usually effective. However, for protein sequences, some substitutions are much more likely than others. The performance of an alignment program is clearly improved when the scoring matrix employed accounts for this difference. As a result, a variety of amino acid properties have been used for constructing substitution matrices in the papers of McLachlan [136], Taylor [188], Rao [142], Overington et al. [156], and Risler et al. [172].

8.1

PAM matrices are due to Dayhoff, Schwartz, and Orcutt [55]. Although the details of Dayhoff's approach had been criticized in the paper of Wilbur [203], PAM matrices remained popular as scoring matrices for protein sequence comparison in the past 30 years. New versions of the Dayhoff matrices were obtained from recalculating the model parameters using more protein sequence data in the papers of Gonnet, Cohen, and Benner [77] and Jones, Taylor, and Thornton [97]. The Dayhoff approach was also extended to estimate a Markov model of amino acid substitution from alignments of remotely homologous sequences by Müller, Spang, and Vingron [147, 148] and Arvestad [16]. For information on the statistical approach to the estimation of substituting matrices, we refer the reader to the survey paper of Yap and Speed [209].

8.2

The BLOSUM matrices are due to Henikoff and Henikoff [88]. They were derived from direct estimation of target frequencies based on relatively distant, presumed correct sequence alignments. Although these matrices do not rely on fitting an evolutionary model, they are the most effective ones for homology search as demonstrated in the papers of Henikoff and Henikoff [89] and Pearson [159].

8.3

Equation (8.2) relating substitution matrices to target frequencies is established by Altschul [3]. The one-to-one correspondence between valid scoring matrices and sets of target frequencies is proved by Yu, Wootton, and Altschul [211].

8.4

Scoring matrix selection is critical for sequence alignment applications. The theoretic information approach to scoring matrix selection is due to Altschul [3]. Henikoff and Henikoff [89] evaluated amino acid scoring matrices using the BLASTP searching program and the PROSITE database. Their study suggests that BLOSUM62 is the best single scoring matrix for database search applications. The empirical study of Pearson [159] illustrates that scaling similarity scores by the logarithm of the size of the database can dramatically improve the performance of a scoring matrix.

8.5

The method discussed in this section is from the paper of Yu and Altschul [212]. An improved method is given in the paper of Altschul et al. [9].

8.6

This section is written based on the paper of States, Gish, and Altschul [184]. The Henikoff and Henikoff method was used to construct scoring matrix for aligning non-coding genomic sequences in the paper of Chiaromonte, Yap, and Miller [45] (see also [46]). More methods for nucleotide scoring matrix can be found in the papers of Müller, Spang, and Vingron [148] and Schwartz et al. [178]. The transition and transversion rate is given in the paper of Li, Wu, and Luo [126].

8.7

The affine gap cost was first proposed by Smith and Waterman [180]. The generalized affine gap cost discussed in this section is due to Altschul [1]. When $c = 2b$, the generalized affine gap cost reduces to a cost model proposed by Zuker and Somorjal for protein structural alignment [217]. The empirical study of Zachariah et al. [213] shows that the generalized affine gap model allows fewer residue pairs aligned than the affine gap model but achieves significantly higher per-residue accuracy. The empirical studies on the distribution of insertions/deletions are found in the papers of Benner, Cohen, and Gonnet [26] and Pascarella and Argos [158].

	A	R	N	D	C	Q	E	G	H	I	L	K	M	F	P	S	T	W	Y	V
A	2																			
R	-2	6																		
N	0	0	2																	
D	0	-1	2	4																
C	-2	-4	-4	-5	12															
Q	0	1	1	2	-5	4														
E	0	-1	1	3	-5	2	4													
G	1	-3	0	1	-3	-1	0	5												
H	-1	2	2	1	-3	3	1	-2	6											
I	-1	-2	-2	-2	-2	-2	-2	-3	-2	5										
L	-2	-3	-3	-4	-6	-2	-3	-4	-2	2	6									
K	-1	3	1	0	-5	1	0	-2	0	-2	-3	5								
M	-1	0	-2	-3	-5	-1	-2	-3	-2	2	4	0	6							
F	-4	-4	-4	-6	-4	-5	-5	-5	-2	1	2	-5	0	9						
P	1	0	-1	-1	-3	0	-1	-1	0	-2	-3	-1	-2	-5	6					
S	1	0	1	0	0	-1	0	1	-1	-1	-3	0	-2	-3	1	2				
T	1	-1	0	0	-2	-1	0	0	-1	0	-2	0	-1	-3	0	1	3			
W	-6	2	-4	-7	-8	-5	-7	-7	-3	-5	-2	-3	-4	0	-6	-2	-5	17		
Y	-3	-4	-2	-4	0	-4	-4	-5	0	-1	-1	-4	-2	7	-5	-3	-3	0	10	
V	0	-2	-2	-2	-2	-2	-2	-1	-2	4	2	-2	2	-1	-1	-1	0	-6	-2	4

Fig. 8.3 The PAM250 matrix.

	A	R	N	D	C	Q	E	G	H	I	L	K	M	F	P	S	T	W	Y	V
A	6																			
R	-7	8																		
N	-4	-6	8																	
D	-3	-10	2	8																
C	-6	-8	-11	-14	10															
Q	-4	-2	-3	-2	-14	8														
E	-2	-9	-2	2	-14	1	8													
G	-2	-9	-3	-3	-9	-7	-4	6												
H	-7	-2	0	-4	-7	1	-5	-9	9											
I	-5	-5	-5	-7	-6	-8	-5	-11	-9	8										
L	-6	-8	-7	-12	-15	-5	-9	-10	-6	-1	7									
K	-7	0	-1	-4	-14	-3	-4	-7	-6	-6	-8	7								
M	-5	-4	-9	-11	-13	-4	-7	-8	-10	-1	1	-2	11							
F	-8	-9	-9	-15	-13	-13	-14	-9	-6	-2	-3	-14	-4	9						
P	-2	-4	-6	-8	-8	-3	-5	-6	-4	-8	-7	-6	-8	-10	8					
S	0	-3	0	-4	-3	-5	-4	-2	-6	-7	-8	-4	-5	-6	-2	6				
T	-1	-6	-2	-5	-8	-5	-6	-6	-7	-2	-7	-3	-4	-9	-4	0	7			
W	-13	-2	-8	-15	-15	-13	-17	-15	-7	-14	-6	-12	-13	-4	-14	-5	-13	13		
Y	-8	-10	-4	-11	-4	-12	-8	-14	-3	-6	-7	-9	-11	2	-13	-7	-6	-5	10	
V	-2	-8	-8	-8	-6	-7	-6	-5	-6	2	-2	-9	-1	-8	-6	-6	-3	-15	-7	7

Fig. 8.4 The PAM30 matrix.

	A	R	N	D	C	Q	E	G	H	I	L	K	M	F	P	S	T	W	Y	V
A	5																			
R	-4	8																		
N	-2	-3	6																	
D	-1	-6	3	6																
C	-4	-5	-7	-9	9															
Q	-2	0	-1	0	-9	7														
E	-1	-5	0	3	-9	2	6													
G	0	-6	-1	-1	-6	-4	-2	6												
H	-4	0	1	-1	-5	2	-2	-6	8											
I	-2	-3	-3	-5	-4	-5	-4	-6	-6	7										
L	-4	-6	-5	-8	-10	-3	-6	-7	-4	1	6									
K	-4	2	0	-2	-9	-1	-2	-5	-3	-4	-5	6								
M	-3	-2	-5	-7	-9	-2	-4	-6	-6	1	2	0	10							
F	-6	-7	-6	-10	-8	-9	-9	-7	-4	0	-1	-9	-2	8						
P	0	-2	-3	-4	-5	-1	-3	-3	-2	-5	-5	-4	-5	-7	7					
S	1	-1	1	-1	-1	-3	-2	-0	-3	-4	-6	-2	-3	-4	0	5				
T	1	-4	0	-2	-5	-3	-3	-3	-4	-1	-4	-1	-2	-6	-2	2	6			
W	-9	0	-6	-10	-11	-8	-11	-10	-5	-9	-4	-7	-8	-2	-9	-3	-8	13		
Y	-5	-7	-3	-7	-2	-8	-6	-9	-1	-4	-4	-7	-7	4	-9	-5	-4	-3	9	
V	-1	-5	-5	-5	-4	-4	-4	-3	-4	3	0	-6	0	-5	-3	-3	-1	-10	-5	6
	A	R	N	D	C	Q	E	G	H	I	L	K	M	F	P	S	T	W	Y	V

Fig. 8.5 The PAM70 matrix.

	A	R	N	D	C	Q	E	G	H	I	L	K	M	F	P	S	T	W	Y	V
A	3																			
R	-3	6																		
N	-1	-1	4																	
D	0	-3	2	5																
C	-3	-4	-5	-7	9															
Q	-1	1	0	1	-7	6														
E	0	-3	1	3	-7	2	5													
G	1	-4	0	0	-4	-3	-1	5												
H	-3	1	2	0	-4	4	-1	-4	7											
I	-1	-2	-2	-3	-3	-3	-3	-4	-4	6										
L	-3	-4	-4	-5	-7	-2	-4	-5	-3	1	5									
K	-2	2	1	-1	-7	0	-1	-3	-2	-3	-4	5								
M	-2	-1	-3	-4	-6	-1	-3	-4	-4	1	3	0	8							
F	-4	-5	-4	-7	-6	-6	-7	-5	-3	0	0	-7	-1	8						
P	1	-1	-2	-3	-4	0	-2	-2	-1	-3	-3	-2	-3	-5	6					
S	1	-1	1	0	0	-2	-1	1	-2	-2	-4	-1	-2	-3	1	3				
T	1	-2	0	-1	-3	-2	-2	-1	-3	0	-3	-1	-1	-4	-1	2	4			
W	-7	1	-4	-8	-8	-6	-8	-8	-3	-6	-3	-5	-6	-1	-7	-2	-6	12		
Y	-4	-5	-2	-5	-1	-5	-5	-6	-1	-2	-2	-5	-4	4	-6	-3	-3	-2	8	
V	0	-3	-3	-3	-3	-3	-3	-2	-3	3	1	-4	1	-3	-2	-2	0	-8	-3	5
	A	R	N	D	C	Q	E	G	H	I	L	K	M	F	P	S	T	W	Y	V

Fig. 8.6 The PAM120 matrix.

A	4																			
R	-1	5																		
N	-2	0	6																	
D	-2	-2	1	6																
C	0	-3	-3	-3	9															
Q	-1	1	0	0	-3	5														
E	-1	0	0	2	-4	2	5													
G	0	-2	0	-1	-3	-2	-2	6												
H	-2	0	1	-1	-3	0	0	-2	8											
I	-1	-3	-3	-3	-1	-3	-3	-4	-3	4										
L	-1	-2	-3	-4	-1	-2	-3	-4	-3	2	4									
K	-1	2	0	-1	-3	1	1	-2	-1	-3	-2	5								
M	-1	-1	-2	-3	-1	0	-2	-3	-2	1	2	-1	5							
F	-2	-3	-3	-3	-2	-3	-3	-3	-1	0	0	-3	0	6						
P	-1	-2	-2	-1	-3	-1	-1	-2	-2	-3	-3	-1	-2	-4	7					
S	1	-1	1	0	-1	0	0	0	-1	-2	-2	0	-1	-2	-1	4				
T	0	-1	0	-1	-1	-1	-1	-2	-2	-1	-1	-1	-1	-2	-1	1	5			
W	-3	-3	-4	-4	-2	-2	-3	-2	-2	-3	-2	-3	-1	1	-4	-3	-2	11		
Y	-2	-2	-2	-3	-2	-1	-2	-3	2	-1	-1	-2	-1	3	-3	-2	-2	2	7	
V	0	-3	-3	-3	-1	-2	-2	-3	-3	3	1	-2	1	-1	-2	-2	0	-3	-1	4
	A	R	N	D	C	Q	E	G	H	I	L	K	M	F	P	S	T	W	Y	V

Fig. 8.7 The BLOSUM62 matrix.

A	5																			
R	-2	7																		
N	-1	0	6																	
D	-2	-1	2	7																
C	-1	-3	-2	-3	12															
Q	-1	1	0	0	-3	6														
E	-1	0	0	2	-3	2	6													
G	0	-2	0	-1	-3	-2	-2	7												
H	-2	0	1	0	-3	1	0	-2	10											
I	-1	-3	-2	-4	-3	-2	-3	-4	-3	5										
L	-1	-2	-3	-3	-2	-2	-2	-3	-2	2	5									
K	-1	3	0	0	-3	1	1	-2	-1	-3	-3	5								
M	-1	-1	-2	-3	-2	0	-2	-2	0	2	2	-1	6							
F	-2	-2	-2	-4	-2	-4	-3	-3	-2	0	1	-3	0	8						
P	-1	-2	-2	-1	-4	-1	0	-2	-2	-2	-3	-1	-2	-3	9					
S	1	-1	1	0	-1	0	0	0	-1	-2	-3	-1	-2	-2	-1	4				
T	0	-1	0	-1	-1	-1	-1	-2	-2	-1	-1	-1	-1	-1	-1	2	5			
W	-2	-2	-4	-4	-5	-2	-3	-2	-3	-2	-2	-2	-2	1	-3	-4	-3	15		
Y	-2	-1	-2	-2	-3	-1	-2	-3	2	0	0	-1	0	3	-3	-2	-1	3	8	
V	0	-2	-3	-3	-1	-3	-3	-3	-3	3	1	-2	1	0	-3	-1	0	-3	-1	5
	A	R	N	D	C	Q	E	G	H	I	L	K	M	F	P	S	T	W	Y	V

Fig. 8.8 The BLOSUM45 matrix.

	A	R	N	D	C	Q	E	G	H	I	L	K	M	F	P	S	T	W	Y	V
A	7																			
R	-3	9																		
N	-3	-1	9																	
D	-3	-3	2	10																
C	-1	-6	-5	-7	13															
Q	-2	1	0	-1	-5	9														
E	-2	-1	-1	2	-7	3	8													
G	0	-4	-1	-3	-6	-4	-4	9												
H	-3	0	1	-2	-7	1	0	-4	12											
I	-3	-5	-6	-7	-2	-5	-6	-7	-6	7										
L	-3	-4	-6	-7	-3	-4	-6	-7	-5	2	6									
K	-1	3	0	-2	-6	2	1	-3	-1	-5	-4	8								
M	-2	-3	-4	-6	-3	-1	-4	-5	-4	2	3	-3	9							
F	-4	-5	-6	-6	-4	-5	-6	-6	-2	-1	0	-5	0	10						
P	-1	-3	-4	-3	-6	-3	-2	-5	-4	-5	-5	-2	-4	-6	12					
S	2	-2	1	-1	-2	-1	-1	-1	-2	-4	-4	-1	-3	-4	-2	7				
T	0	-2	0	-2	-2	-1	-2	-3	-3	-2	-3	-1	-1	-4	-3	2	8			
W	-5	-5	-7	-8	-5	-4	-6	-6	-4	-5	-4	-6	-3	0	-7	-6	-5	16		
Y	-4	-4	-4	-6	-5	-3	-5	-6	3	-3	-2	-4	-3	4	-6	-3	-3	3	11	
V	-1	-4	-5	-6	-2	-4	-4	-6	-5	4	1	-4	1	-2	-4	-3	0	-5	-3	7

Fig. 8.9 The BLOSUM80 matrix.

Appendix A
Basic Concepts in Molecular Biology

Deoxyribonucleic acid (DNA) is the genetic material of living organisms. It carries information in a coded form from parent to offspring and from cell to cell. A gene is a linear array of nucleotides located in a particular position on a particular chromosome that encodes a specific functional product (a protein or RNA molecule). When a gene is activated, its information is transcribed first into another nucleic acid, ribonucleic acid (RNA), which in turn directs the synthesis of the gene products, the specific proteins. This appendix introduces some basic concepts of DNA, proteins, genes, and genomes.

A.1 The Nucleic Acids: DNA and RNA

DNA is made up of four chemicals – adenine, cytosine, guanine, and thymine – that occur millions or billions of times throughout a genome. The human genome, for example, has about three billion pairs of bases. RNA is made of four chemicals: adenine, cytosine, guanine, and uracil. The bases are usually referred to by their initial letters: A, C, G, T for DNA and A, C, G, U for RNA.

The particular order of As, Cs, Gs, and Ts is extremely important. The order underlies all of life's diversity. It even determines whether an organism is human or another species such as yeast, fruit fly, or chimpanzee, all of which have their own genomes.

In the late 1940s, Erwin Chargaff noted an important similarity: the amount of adenine in DNA molecules is always equal to the amount of thymine, and the amount of guanine is always equal to the amount of cytosine (#A = #T and #G = #C). In 1953, based on the x-ray diffraction data of Rosalind Franklin and Maurice Wilkins, James Watson and Francis Crick proposed a conceptual model for DNA structure. The Watson-Crick model states that the DNA molecule is a double helix, in which two strands are twisted together. The only two possible pairs are AT and CG. This yields a molecule in which #A = #T and #G = #C. The model also suggests that the basis for copying the genetic information is the complementarity of its

Table A.1 Twenty different amino acids and their abbreviations.

Amino Acid	3-Letter	1-Letter
Alanine	Ala	A
Arginine	Arg	R
Asparagine	Asn	N
Aspartic acid	Asp	D
Cysteine	Cys	C
Glutamic acid	Glu	E
Glutamine	Gln	Q
Glycine	Gly	G
Histidine	His	H
Isoleucine	Ile	I
Leucine	Leu	L
Lysine	Lys	K
Methionine	Met	M
Phenylalanine	Phe	F
Proline	Pro	P
Serine	Ser	S
Threonine	Thr	T
Tryptophan	Trp	W
Tyrosine	Tyr	Y
Valine	Val	V

bases. For example, if the sequence on one strand is AGATC, then the sequence of
the other strand would have to be TCTAG – its complementary bases.

There are several different kinds of RNA made by the cell. In particular, mRNA,
messenger RNA, is a copy of a gene. It acts as a photocopy of a gene by having a
sequence complementary to one strand of the DNA and identical to the other strand.
Other RNAs include tRNA (transfer RNA), rRNA (ribosomal RNA), and snRNA
(small nuclear RNA). Since RNA cannot form a stable double helix, it actually exists
as a single-stranded molecule. However, some regions can form hairpin loops if
there is some base pair complementation (A and U, C and G). The RNA molecule
with its hairpin loops is said to have a secondary structure.

A.2 Proteins

The building blocks of proteins are the amino acids. Only 20 different amino acids
make up the diverse array of proteins found in living organisms. Table A.1 summarizes these 20 common amino acids. Each protein differs according to the amount,
type and arrangement of amino acids that make up its structure. The chains of amino
acids are linked by peptide bonds. A long chain of amino acids linked by peptide
bonds is a polypeptide. Proteins are long, complex polypeptides.

The sequence of amino acids that makes up a particular polypeptide chain is
called the primary structure of a protein. The primary structure folds into the sec-

ondary structure, which is the path that the polypeptide backbone of the protein follows in space. The tertiary structure is the organization in three dimensions of all the atoms in the polypeptide chain. The quaternary structure consists of aggregates of more than one polypeptide chain. The structure of a protein is crucial to its functionality.

A.3 Genes

Genes are the fundamental physical and functional units of heredity. Genes carry information for making all the proteins required by all organisms, which in turn determine how the organism functions.

A gene is an ordered sequence of nucleotides located in a particular position on a particular chromosome that encodes a specific functional product (a protein or RNA molecule). Expressed genes include those that are transcribed into mRNA and then translated into protein and those that are transcribed into RNA but not translated into protein (*e.g.*, tRNA and rRNA).

How does a segment of a strand of DNA relate to the production of the amino acid sequence of a protein? This concept is well explained by the *central dogma of molecular biology*. Information flow (with the exception of reverse transcription) is from DNA to RNA via the process of transcription, and then to protein via translation. Transcription is the making of an RNA molecule off a DNA template. Translation is the construction of an amino acid sequence (polypeptide) from an RNA molecule.

How does an mRNA template specify amino acid sequence? The answer lies in the genetic code. It would be impossible for each amino acid to be specified by one or two nucleotides, because there are 20 types of amino acids, and yet there are only four types of nucleotides. Indeed, each amino acid is specified by a particular combination of three nucleotides, called a codon, which is a triplet on the mRNA that codes for either a specific amino acid or a control word. The genetic code was broken by Marshall Nirenberg and Heinrich Matthaei a decade after Watson and Crick's work.

A.4 The Genomes

A genome is all the DNA in an organism, including its genes.[1] In 1990, the Human Genome Project was launched by the U.S. Department of Energy and the National Institutes of Health. The project originally was planned to last 15 years, but rapid technological advances accelerated the draft completion date to 2003. The goals of the project were to identify all the genes in human DNA, determine the

---

[1] Wouldn't you agree that the genomes are the largest programs written in the oldest language and are quite adaptable, flexible, and fault-tolerant?

sequences of the 3 billion base pairs that make up human DNA, store this information in databases, develop tools for data analysis, and address the ethical, legal, and social issues that may arise from the project.

Because all organisms are related through similarities in DNA sequences, insights gained from other organisms often provide the basis for comparative studies that are critical to understanding more complex biological systems. As the sequencing technology advances, digital genomes for many species are now available for researchers to query and compare. Interested readers are encouraged to check out the latest update at http://www.ncbi.nlm.nih.gov/sites/entrez?db=genomeprj.

Appendix B
Elementary Probability Theory

This chapter summaries basic background knowledge and establishes the book's terminology and notation in probability theory. We give informal definitions for events, random variables, and probability distributions and list statements without proof. The reader is referred to consult an elementary probability textbook for the need of justification.

B.1 Events and Probabilities

Intuitively, an *event* is something that will or will not occur in a situation depending on chance. Such a situation is called an *experiment*. For example, in the experiment of rolling a die, an event might be that the number turning up is 5. In an alignment of two DNA sequences, whether the two sequences have the same nucleotide at a position is an event.

The *certain* event, denoted by Ω, always occurs. The *impossible* event, denoted by ϕ, never occurs. In different contexts these two events might be described in different, but essentially identical ways.

Let A be an event. The *complementary* event of A is the event that A does not occur and is written $\bar{A}$.

Let A_1 and A_2 be two events. The event that at least one of A_1 or A_2 occurs is called the *union* of A_1 and A_2 and is written $A_1 \cup A_2$. The event that both A_1 and A_2 occur is called the *intersection* of A_1 and A_2 and is written $A_1 \cap A_2$. We say A_1 and A_2 are *disjoint*, or *mutually exclusive*, if they cannot occur together. In this case, $A_1 \cap A_2 = \phi$. These definitions extend to finite number of events in a natural way. The union of the events $A_1, A_2, \ldots, A_n$ is the event that at least one of these events occurs and is written $A_1 \cup A_2 \cup \cdots \cup A_n$ or $\cup_{i=1}^n A_i$. The intersection of the events $A_1, A_2, \ldots, A_n$ is the event that all of these events occur, and is written $A_1 \cap A_2 \cap \cdots \cap A_n$, or simply $A_1 A_2 \cdots A_n$.

The probability of an event A is written $Pr[A]$. Obviously, for the certain event Ω, $\Pr[\Omega] = 1$; for the impossible event ϕ, $\Pr[\phi] = 0$. For disjoint events A_1 and A_2,

$\Pr[A_1 \cup A_2] = \Pr[A_1] + \Pr[A_2]$. In general, for events $A_1, A_2, \ldots, A_n$ such that A_i and A_j disjoint whenever $i \neq j$, $\Pr[\cup_{i=1}^n A_i] = \sum_{i=1}^n \Pr[A_i]$. Let $A_1, A_2, \ldots, A_n$ be disjoint events such that exactly one of these events will occur. Then,

$$\Pr[B] = \sum_{i=1}^n \Pr[A_i B] \tag{B.1}$$

for any event B. This formula is called the *law of total probability*.

Events A_1 and A_2 are *independent* if $\Pr[A_1 \cap A_2] = \Pr[A_1]\Pr[A_2]$. In general, events $A_1, A_2, \ldots, A_n$ are *independent* if

$$\Pr[A_{i_1} \cap A_{i_2} \cap \cdots \cap A_{i_k}] = Pr[A_{i_1}]\Pr[A_{i_2}] \cdots \Pr[A_{i_k}]$$

for any subset of distinct indices $i_1, i_2, \ldots, i_k$.

Let A and B be two events with $\Pr[A] \neq 0$. The conditional probability that B occurs given that A occurs is written $\Pr[B|A]$ and defined by

$$\Pr[B|A] = \frac{\Pr[AB]}{\Pr[A]}. \tag{B.2}$$

This definition has several important consequences. First, it is equivalent to

$$\Pr[AB] = \Pr[A]\Pr[B|A]. \tag{B.3}$$

Combining (B.3) and the law of total probability (B.1), we obtain

$$\Pr[B] = \sum_{i=1}^n \Pr[A_i]\Pr[B|A_i]. \tag{B.4}$$

Another consequence of the conditional probability is

$$\Pr[A|B] = \frac{\Pr[A]\Pr[B|A]}{\Pr[B]} \tag{B.5}$$

when $\Pr[A] \neq 0$ and $\Pr[B] \neq 0$. This is called Bayes' formula in Bayesian statistics.

B.2 Random Variables

A *random* variable is a variable that takes its value by chance. Random variables are often written as capital letters such as X, Y, and Z, whereas the observed values of a random variable are written in lowercase letters such as x, y, and z. For a random variable X and a real number x, the expression $\{X \leq x\}$ denotes the event that X has a value less than or equal to x. The probability that the event occurs is denoted by $\Pr[X \leq x]$.

A random variable is *discrete* if it takes a value from a discrete set of numbers. For example, the number of heads turning up in the experiment of tossing a coin n times is a discrete random variable. In this case, the possible values of the random variable are $0, 1, 2, \ldots, n$. The *probability distribution* of a random variable is often presented in the form of a table, a chart, or a mathematical function. The *distribution function* of a discrete random variable X is written $F_X(x)$ and defined as

$$F_X(x) = \sum_{v \leq x} \Pr[X = v], \quad -\infty < x < \infty. \tag{B.6}$$

Notice that F_X is a step function for a discrete random variable X.

A random variable is *continuous* if it takes any value from a continuous interval. The probability for a continuous random variable is not allocated to specific values, but rather to intervals of values. If there is a nonnegative function $f(x)$ defined for $-\infty < x < \infty$ such that

$$\Pr[a < X \leq b] = \int_a^b f(x)dx, \quad \infty < a < b < \infty,$$

then $f(x)$ is called the *probability density function* of the random variable X. If X has a probability density function $f_X(x)$, then its *distribution function* $F_X(x)$ can be written as

$$F_X(x) = \Pr[X \leq x] = \int_{-\infty}^x f_X(x), \quad -\infty < x < \infty. \tag{B.7}$$

B.3 Major Discrete Distributions

In this section, we simply list the important discrete probability distributions that appear frequently in bioinformatics.

B.3.1 Bernoulli Distribution

A Bernoulli trial is a single experiment with two possible outcomes "success" and "failure." The Bernoulli random variable X associated with a Bernoulli trial takes only two possible values 0 and 1 with the following probability distribution:

$$\Pr[X = x] = p^x(1 - p)^{1-x}, \quad x = 0, 1, \tag{B.8}$$

where p is called the *parameter* of X.

In probability theory, Bernoulli random variables occur often as indicators of events. The *indicator* I_A of an event A is the random variable that has 1 if A occurs and 0 otherwise. The I_A is a Bernoulli random variable with parameter $\Pr[A]$.

B.3.2 Binomial Distribution

A binomial random variable X is the number of successes in a fixed number n of independent Bernoulli trials with parameter p. The p and n are called the *parameter* and *index* of X, respectively. In particular, the Bernoulli trial can be considered as a binomial distribution with index 1.

The probability distribution of the binomial random variable X with index n and parameter p is

$$\Pr[X = k] = \frac{n!}{k!(n-k)!} p^k (1-p)^{n-k}, \quad k = 0, 1, \ldots, n. \tag{B.9}$$

B.3.3 Geometric and Geometric-like Distributions

A random variable X has a geometric distribution with *parameter p* if

$$\Pr[X = k] = (1-p)p^k, \quad k = 0, 1, 2, \ldots. \tag{B.10}$$

The geometric distribution arises from independent Bernoulli trials. Suppose that a sequence of independent Bernoulli trials are conducted, each trial having probability p of success. The number of successes prior to the first failure has the geometric distribution with parameter p.

By (B.10), the distribution function of the geometric random variable X with parameter p is

$$F_X(x) = 1 - p^{x+1}, \quad x = 0, 1, 2, \ldots. \tag{B.11}$$

Suppose that Y is a random variable taking possible values $0, 1, 2, \ldots$ and $0 < p < 1$. If

$$\lim_{y \to \infty} \frac{1 - F_Y(y)}{p^{y+1}} = C$$

for some fixed C, then the random variable Y is said to be *geometric-like*. Geometric-like distributions play a central role in BLAST statistic theory studied in Chapter 7.

B.3.4 The Poisson Distribution

A random variable X has the Poisson distribution with *parameter λ* if

$$\Pr[X = x] = \frac{1}{x!} e^{-\lambda} \lambda^x, \quad x = 0, 1, 2, \ldots. \tag{B.12}$$

The Poisson distribution is probably the most important discrete distribution. It not only has elegant mathematical properties but also is thought of as the law of the rare events. For example, in a binomial distribution, if the number of trials n is large and the probability of success p for each trial is small such that $\lambda = np$ remains constant, then the binomial distribution converges to the Poisson distribution with parameter λ.

Applying integration by part, we see that the distribution function X of the Poisson distribution with parameter λ is

$$F_X(k) = \Pr[X \leq k] = \frac{\lambda^{k+1}}{k!} \int_1^\infty y^k e^{-\lambda y} dy, \quad k = 0, 1, 2, \ldots. \tag{B.13}$$

B.3.5 Probability Generating Function

For a discrete random variable X, its probability generating function (pgf) is written $G(t)$ and defined as

$$G(t) = \sum_{x \in I} \Pr[X = x] t^x, \tag{B.14}$$

where I is the set of all possible values of X. This sum function always converges to 1 for $t = 1$ and often converges in an open interval containing 1 for all probability distributions of interest to us. Probability generating functions have the following two basic properties.

First, probability generating functions have one-to-one relationship with probability distributions. Knowing the pgf is equivalent to knowing the probability distribution for a random variable to certain extent. In particular, for a nonnegative integer-valued random variable X, we have

$$\Pr[X = k] = \frac{1}{k!} \left. \frac{d^k G(t)}{dt^k} \right|_{t=0}. \tag{B.15}$$

Second, if random variables $X_1, X_2, \ldots, X_n$ are independent and have pgfs $G_1(t)$, $G_2(t), \ldots, G_n(t)$ respectively, then the pgf $G(t)$ of their sum

$$X = X_1 + X_2 + \cdots + X_n$$

is simply the product

$$G(t) = G_1(t) G_2(t) \cdots G_n(t). \tag{B.16}$$

B.4 Major Continuous Distributions

In this section, we list several important continuous distributions and their simple properties.

B.4.1 Uniform Distribution

A continuous random variable X has the uniform distribution over an interval $[a,b]$, $a < b$, if it has the density function

$$f_X(x) = \frac{1}{b-a}, \quad a \leq x \leq b \qquad (B.17)$$

and hence the distribution function

$$F_X(x) = \frac{x-a}{b-a}, \quad a \leq x \leq b. \qquad (B.18)$$

B.4.2 Exponential Distribution

A continuous random variable X has the exponential distribution with parameter λ if its density function is

$$f_X(x) = \lambda e^{-\lambda x}, \quad x \geq 0. \qquad (B.19)$$

By integration, the distribution function is

$$F_X(x) = \Pr[X \leq x] = 1 - e^{-\lambda x}, \quad x \geq 0. \qquad (B.20)$$

The exponential distribution is the continuous analogue of the geometric distribution. Suppose that a random variable X has the exponential distribution with parameter λ. We define

$$Y = \lceil X \rceil.$$

Y is a discrete random variable having possible values $0, 1, 2, \ldots$. Moreover, by (B.19),

$$\Pr[Y = k] = \Pr[k \leq X < k+1] = \int_k^{k+1} f_X(x)dx = (1 - e^{-\lambda})e^{-\lambda k}, \quad k = 0, 1, \ldots.$$

This shows that Y has the geometric distribution with parameter $e^{-\lambda}$.

Applying the definition of conditional probability (B.2), we obtain, for $x, t > 0$,

$$\Pr[X > t + x | X > t]$$
$$= \Pr[X > t + x]/\Pr[X > t]$$
$$= e^{-\lambda x}$$
$$= \Pr[X > x].$$

Therefore, we often say that the exponential distribution has the *memoryless prop-erty*.

B.4.3 Normal Distribution

A continuous random variable X has the normal distribution with parameters μ and $\sigma > 0$ if it has density function

$$f_X(x) = \frac{1}{\sqrt{2\pi}\sigma} e^{-(x-\mu)^2/(2\sigma^2)}, \quad -\infty < x < \infty. \tag{B.21}$$

The function f_X is bell-shaped and symmetric about $x = \mu$. By convention, such a random variable is called an $N(\mu, \sigma^2)$ random variable.

The normal distribution with parameters 0 and 1 is called the *standard* normal distribution. Suppose that X is an $N(\mu, \sigma^2)$ random variable. Then, the random variable Z defined by $Z = \frac{X-\mu}{\sigma}$ has the standard normal distribution. Because of this, the value of Z is often called a *z*-score: If x is the observed value of X, then its *z*-score is $\frac{x-\mu}{\sigma}$.

Let a random variable X have a normal distribution. For arbitrary a and b, $a < b$, we cannot find the probability $\Pr[a < X < b]$ by simply integrating its density function in closed form. Instead, we reduce it to a probability statement for the standard normal distribution and find an accurate approximation of it from tables that are widely available in a textbook in probability theory.

B.5 Mean, Variance, and Moments

B.5.1 The Mean of a Random Variable

The mean of a discrete random variable X is written μ_X and defined as

$$\mu_X = \sum_{x \in I} x \Pr[X = x] \tag{B.22}$$

where I is the range of X provided that the sum converges. The mean μ_X is also called the expected value of X and hence often written as $E(X)$. If X has the pgf $G(t)$, then the mean of X is equal to the derivative of $G(t)$ at $t = 1$. That is,

Table B.1 Means and variances of major probability distributions.

Distribution	Parameters	Mean	Variance
Bernoulli	p	p	$p(1-p)$
Binomial	p,n	np	$np(1-p)$
Geometric	p	$p/1-p$	$p/(1-p)^2$
Poisson	λ	λ	λ
Uniform	$[a,b]$	$(a+b)/2$	$(b-a)^2/12$
Exponential	λ	$1/\lambda$	$1/\lambda^2$
Normal	μ, σ^2	μ	σ^2

$$\mu_X = \left.\frac{dG(t)}{dt}\right|_{t=1} \tag{B.23}$$

If X is a discrete random variable with range I and g is a function on I, then $g(X)$ is also a discrete random variable. The expected value of $g(X)$ is written $E(g(X))$ and defined as

$$E(g(X)) = \sum_{x \in I} g(x)\Pr[X=x]. \tag{B.24}$$

Similarly, for a continuous random variable Y having density function f_Y, its mean is defined as

$$\mu_Y = \int_{-\infty}^{\infty} y f_Y(y)d(y). \tag{B.25}$$

If g is a function, the expected value of $g(Y)$ is defined as

$$E(g(Y)) = \int_{-\infty}^{\infty} g(y) f_Y(y)d(y). \tag{B.26}$$

The means of random variables with a distribution described in Sections B.3 and B.4 are listed in Table B.1.

Let $X_1, X_2, \dots, X_n$ be random variables having the same range I. Then, for any constants $a_1, a_2, \dots, a_n$,

$$E\left(\sum_{i=1}^{n} a_i X_i\right) = \sum_{i=1}^{n} a_i E(X_i). \tag{B.27}$$

This is called the linearity property of the mean.

For any positive integer, $E(X^m)$ is called the rth *moment* of a random variable X provided that the sum exists.

B.5.2 The Variance of a Random Variable

Another important quantity of a random variable is its variance. The *variance* of a discrete random variable X is written σ_X^2 or $\text{Var}(X)$ and defined as

$$\sigma_X^2 = \sum_{x \in I} (x - \mu_X)^2 \Pr[X = x] \qquad (\text{B.28})$$

where I is the range of X. We can verify that

$$\sigma_X^2 = E(X^2) - \mu_X^2.$$

Similarly, the variance of a continuous random variable Y is defined as

$$\sigma_Y^2 = \int_{-\infty}^{\infty} (y - \mu_Y)^2 f_Y(y) dy. \qquad (\text{B.29})$$

The variance of random variables with a distribution described in Sections B.3 and B.4 are listed in Table B.1.

The square root of the variance is called the *standard deviation*. Notice that $\sigma_X^2 = E((X - \mu_X)^2)$. In general, for a random variable X and any positive integer r, $E((X - \mu_X)^r)$ is called the rth *central moment* of X provided that it exists.

For any positive ε, by definition,

$$\sigma^2 = E((X - \mu_X)^2) \geq \varepsilon^2 \sigma^2 \Pr[|X - \mu_X| \geq \varepsilon \sigma],$$

or equivalently,

$$\Pr[|X - \mu_X| \geq \varepsilon \sigma] \leq \frac{1}{\varepsilon^2}. \qquad (\text{B.30})$$

This inequality is called *Chebyshev's inequality*. It is very useful because it applies to random variables of any distribution. The use of the Chebyshev's inequality is called the second moment method.

Let $X_1, X_2, \ldots, X_n$ be n random variables having the same range I and let

$$X = X_1 + X_2 + \cdots + X_n.$$

Then

$$\sigma_X^2 = \sum_{i=1}^{n} \sigma_{X_i}^2 + \sum_{i \neq j} \text{cov}[X_i, X_j], \qquad (\text{B.31})$$

where $\text{cov}[X_i, X_j] = E[X_i X_j] - E[X_i] E[X_j]$, called the *covariance* of X_i and X_j. Notice that $\text{cov}[X_i, X_j] = 0$ if X_i and X_j are independent.

B.5.3 The Moment-Generating Function

An important technique for studying the moments of a random variable is to use its moment-generating function. The *moment-generating function* (mgf) of a random variable X is written $M_X(t)$ and defined as

$$M_X(t) = E(e^{tX}).\qquad(B.32)$$

For a discrete random variable X having range I,

$$M_X(t) = \sum_{x \in I} e^{tx} \Pr[X = x].$$

For a continuous random variable Y having density function f_Y,

$$M_Y(t) = \int_{-\infty}^{\infty} e^{ty} f_Y(y) d(y).$$

Clearly, the mgf converges to 1 when $t = 0$. It usually converges in an open interval $(-\varepsilon, \varepsilon)$ containing 0.

Any moments of a random variable can be expressed as an appropriate differentiation of its mgf. In particular, for any random variable X,

$$\mu_X = \left(\frac{d\, M_X(t)}{d\, t}\right)_{t=0},\qquad(B.33)$$

and

$$\sigma_X^2 = \left(\frac{d^2\, M_X(t)}{d\, t^2}\right)_{t=0} - \mu_X^2 = \left(\frac{d^2\, \log M_X(t)}{d\, t^2}\right)_{t=0}.\qquad(B.34)$$

Hence, a mgf is useful when it has a simple form. For a random variable having a probability distribution listed in Sections B.3 and B.4, its mgf can be evaluated as a simple form. However, this is not always true.

The following property of mgfs plays a critical role in the scoring matrix and BLAST statistic theory.

Theorem B.1. *Let X be a discrete random variable with $M_X(t)$ converging for all t in $(-\infty, \infty)$. Suppose that X takes at least one negative value and one positive value with nonzero probability and that $E(X) \neq 0$. Then, there exists a unique nonzero value θ such that $M_X(\theta) = 1$.*

Proof. Let $a, b > 0$ such that $\Pr[X = -a] > 0$ and $\Pr[X = b] > 0$. By definition,

$$M_X(t) > \Pr[X = -a] e^{-at}$$

and

$$M_X(t) > \Pr[X = b]e^{bt}$$

Hence $M_X(t) \to +\infty$ as $|t| \to \infty$. Furthermore, because, by (B.34),

$$\frac{d^2 M_X(t)}{dt^2} = \sigma_X^2 + \mu_X^2 > 0,$$

the mgf $M_X(t)$ is convex in $(-\infty, +\infty)$. Recall that $M_X(0) = 1$ and, by (B.33),

$$\left(\frac{d M_X(t)}{dt} \right)_{t=0} = E(X).$$

If $E(X) < 0$, then there exists $\delta > 0$ such that $M_X(\delta) < M_X(0) = 1$. Because $M_X(t)$ is continuous, there exists θ in the interval $(\delta, +\infty)$ such that $M_X(\theta) = 1$.

If $E(X) > 0$, then there exists $\delta < 0$ such that $M_X(\delta) < M_X(0) = 1$. In this case, there exists θ in the interval $(-\infty, \delta)$ such that $M_X(\theta) = 1$. □

Finally, we remark that if the mgf of a random variable converges in an open interval containing 1, then the mgf uniquely determines the probability distribution of the random variable.

B.6 Relative Entropy of Probability Distributions

Suppose that there are two probability distributions P_1 and P_2 having the same range I. The *relative entropy* of P_2 with respect to P_1 is defined by

$$H(P_2, P_1) = \sum_{x \in I} \Pr[X_2 = x] \log \frac{\Pr[X_2 = x]}{\Pr[X_1 = x]} \tag{B.35}$$

where X_i is a random variable having probability distribution P_i for $i = 1, 2$.

Although the relative entropy $H(P_2, P_1)$ measures in some sense the dissimilarity between them, it is not symmetric. Define symmetric function

$$J(P_1, P_2) = H(P_1, P_2) + H(P2, P_1).$$

Because $x - y$ has the same sign as $\log(x/y)$ for any $x, y > 0$,

$$J(P_1, P_2) = \sum_{x \in I} (\Pr[X_2 = x] - \Pr[X_1 = x]) \left(\log \frac{\Pr[X_2 = x]}{\Pr[X_1 = x]} \right) \geq 0,$$

and the equality holds only if $P_1 = P_2$. In information theory literature, $J(P_1, P_2)$ is called the *divergence* between P_1 and P_2.

B.7 Discrete-time Finite Markov Chains

B.7.1 Basic Definitions

A *discrete-time finite Markov chain* X is a stochastic process whose state space is finite and whose time set is $\{0,1,2,\ldots\}$ satisfying the Markov property. At each time point, the Markov process occupies one state in the state space; In each time step from t to $t+1$, the process either does not move or moves to some other state with some probability. The Markov property is that

$$\Pr[X_{t+1} = E \mid X_0 = E_0,\ldots,X_{t-1} = E_{t-1}, X_t = E_t]$$
$$= \Pr[X_{t+1} = E \mid X_t = E_t] \tag{B.36}$$

for each time point t and all states $E, E_0, E_1, \ldots, E_{t-1}$. In other words, the Markov chains have the following characteristics:

(i) The *memoryless* property. If at some time point t the process is in state E', then the probability that at time point $t+1$ it is in state E'' depends only on state E', and not on how the process had reached state E' before time t.

(ii) The *time homogeneity property*. The probability that the process moves from state E' to state E'' in time step t to $t+1$ is independent of t.

The importance of Markov chains lies in that many natural phenomena in physics, biology, and economics can be described by them and is enhanced by the amenability of Markov chains to quantitative analysis.

Let p_{ij} be the probability that the process X moves from E_i to E_j in each time step. The p_{ij}s form the so-called *transition (probability) matrix* of X. We denote the matrix by P_X and write it as

$$P_X = \begin{pmatrix} p_{11} & p_{12} & p_{13} & \cdots & p_{1n} \\ p_{21} & p_{22} & p_{23} & \cdots & p_{2n} \\ \vdots & \vdots & \vdots & \ddots & \vdots \\ p_{n1} & p_{n2} & p_{n3} & \cdots & p_{nn} \end{pmatrix}, \tag{B.37}$$

where n is the number of states in the state space. The rows of P_X corresponds one-to-one to the states respectively. The entries in any particular row are the probabilities that the process moves from the corresponding state in a time step and hence must sum to 1. If the row corresponding to state E has 1 in the diagonal entry, E is an *absorbing* state in the sense that X will never leave E once it enters E.

We denote the probability that X moves from E_i to E_j in k time steps by $p_{ij}^{(k)}$ and set $P_X^{(k)} = \left(p_{ij}^{(k)} \right)$. If X is in E_i at time point t and moves to E_j at time point $t+k$, then it must be in some state E_s at time point $t+k-1$. Therefore,

$$p_{ij}^{(k)} = \sum_s p_{is}^{(k-1)} p_{sj},$$

or equivalently

$$P_X^{(k)} = P_X^{(k-1)} \times P_X = P_X^k. \tag{B.38}$$

It is often assumed that there is some probability π_i that X is in the state E_i at the time 0. The π_is form the *initial probability distribution* of X. The analysis of a Markov chain process is mainly based on the calculation of the probabilities of the possible realizations of the process. Therefore, a Markov chain is completely defined by its initial probability distribution and transition matrix.

A Markov chain is *aperiodic* if there is no state such that the process moves to the state only at $a, 2a, \dots$ steps later for some integer a exceeding 1.

A Markov chain is *irreducible* if any state can be reached from any other state after certain steps. A state is an *absorbing state* if once it is entered, it will never be left. Obviously, an irreducible Markov chain process does not have absorbing states.

B.7.2 Markov Chains with No Absorbing States

In this section, we assume that the Markov chain of interest is finite, aperiodic, and irreducible. Suppose that a Markov chain has transition probability matrix (p_{ij}) over the n states and that at time point t the probability that the process is in state E_i is ϕ_i for $i = 1, 2, \dots, n$. Then

$$\sum_i \phi_i = 1.$$

From the general conditional probability formula (B.4), at time point $t + 1$ the probability that process is in state E_i becomes $\sum_k \phi_k p_{ki}$, $i = 1, 2, \dots, n$. If for every i

$$\phi_i = \sum_k \phi_k p_{ki}, \tag{B.39}$$

the probability that the process is in a state will never change. Hence, the probability distribution satisfying (B.39) is called the *stationary distribution* of the process.

Because the probabilities in each row of P sum to 1, 1 is an eigenvalue of P. The condition (B.39) can be rewritten as

$$(\phi_1, \phi_2, \cdots, \phi_n) = (\phi_1, \phi_2, \cdots, \phi_n) P$$

and hence implies that the vector $(\phi_1, \phi_2, \cdots, \phi_n)$ is the left eigenvector of P corresponding to the eigenvalue 1. Moreover, $(1, 1, \cdots, 1)$ is the right eigenvector of P corresponding to 1. Because the Markov chain is aperiodic and irreducible, all other eigenvalues of P have absolute value less than 1. This gives us that as n increases, P^n approaches the matrix

$$\begin{pmatrix} 1 \\ 1 \\ \vdots \\ 1 \end{pmatrix} (\phi_1 \quad \phi_2 \quad \cdots \quad \phi_n) = \begin{pmatrix} \phi_1 & \phi_2 & \cdots & \phi_n \\ \phi_1 & \phi_2 & \cdots & \phi_n \\ \vdots & \vdots & \ddots & \vdots \\ \phi_1 & \phi_2 & \cdots & \phi_n \end{pmatrix}.$$

One implication of this statement is as follows. No matter what is the initial probability distribution of the starting state, the process will be in state E_i with probability close to ϕ_i after enough steps.

B.7.3 Markov Chains with Absorbing States

Analysis of a Markov chain with absorbing states is often reduced to the following two questions:

(i) What is the probability that the process moves into a particular absorbing state?

(ii) What is the mean time until the process moves into some absorbing state?

These two basic problems can be answered by applying the so-called *first step analysis* method. Suppose that X is a Markov chain with n states $E_1, E_2, \ldots, E_n$ and the first m states are absorbing states, where $m \geq 1$. For $j \leq m$, the probability that X eventually enters E_j rather than other absorbing states depends on the initial state X_0. Let

$$u_{ij} = \Pr[X \text{ enters } E_j \,|X_0 = E_i], \ 1 \leq j \leq m, \ m+1 \leq i \leq n. \qquad (B.40)$$

After the first step, X moves from E_i to state E_k with probability p_{ik}. By the law of total probability (B.4), we have

$$u_{ij} = p_{ij} + \sum_{k>m} p_{ik} u_{kj}, 1 \leq j \leq m, \ m+1 \leq i \leq n. \qquad (B.41)$$

Solving this difference equation usually gives an answer to the first basic question.

To answer the second basic question, we let

$$\mu_i = E[\,X \text{ enters an absorbing state } |X_0 = E_i], \ m+1 \leq i \leq n.$$

After step 1, if X_1 is an absorbing state, then no further steps are required. If, on the other hand, X_1 is a non-absorbing state E_k, then the process is back at its starting point, and, on average, μ_k additional steps are required for entering into absorbing state. Weight these possibilities by their respective probabilities, we have

$$\mu_i = 1 + \sum_{k>m} p_{ik} \mu_k, \ m+1 \leq i \leq n. \qquad (B.42)$$

B.7.4 Random Walks

Random walks are special cases of Markov chain processes. A simple random walk is just a Markov chain process whose state space is a finite or infinite subset of consecutive integers: $0, 1, \ldots, c$, in which the process, if it is in state k, can either stay in k or move to one of the neighboring states $k-1$ and $k+1$. The states 0 and c are often absorbing states.

A classic example of random walk process is about a gambler. Suppose a gambler having initially k dollars plays a series of games. For each game, he has probability p of winning one dollar and probability $1 - p = q$ of losing one dollar. The money Y_n that he has after n games is a random walk process.

In a general random walk, the process may stay or move to one of $m > 2$ nearby states. Although random walks can be analyzed as Markov chains, the special features of a random walk allow simple methods of analysis. For example, the moment-generating approach is a powerful tool for analysis of simple random walks.

B.7.5 High-Order Markov Chains

Markov chain processes satisfy the memoryless and time homogeneity properties. In bioinformatics, more general Markov chain processes are used to model gene-coding sequences. High-order Markov chains relax the memoryless property. A discrete stochastic process is a kth-order Markov chain if

$$\Pr[X_{t+1} = E \mid X_0 = E_0, \cdots, X_{t-1} = E_{t-1}, X_t = E_t]$$
$$= \Pr[X_{t+1} = E | X_t = E_t, X_{t-1} = E_{t-1}, \cdots, X_{t-k+1} = E_{t-k+1}] \qquad \text{(B.43)}$$

for each time point t and all states $E, E_0, E_1, \ldots, E_{t-1}$. In other words, the probability that the process is in a state at the next time point depends on the last k states of the past history.

It is not hard to see that a kth-order Markov chain is completely defined by the initial distribution and the transition probabilities of the form (B.43).

B.8 Recurrent Events and the Renewal Theorem

The discrete renewal theory is a major branch of classical probability theory. It concerns recurrent events occurring in repeated trials. In this section, we state two basic theorem in the renewal theory. The interested reader is referred to the book [68] of Feller for their proofs.

Consider an infinite sequence of repeated trials with possible outcomes X_i ($i = 1, 2, \ldots$). Let E be an event defined by an attribute of finite sequences of possible outcomes X_i. We say that E occurs at the ith trial if the outcomes $X_1, X_2, \ldots, X_i$

have the attribute. E is a *recurrent event* if, under the condition that E occurs at the ith trial, $\Pr[X_1, X_2, \ldots, X_{i+k}$ have the attribute] is equal to the product of $\Pr[X_1, X_2, \ldots, X_i$ have the attribute] and $\Pr[X_{i+1}, X_{i+2}, \ldots, X_{i+k}$ have the attribute]. For instance, the event that a success followed by failure is a recurrent event in a sequence of Bernoulli trials.

For a recurrent event E, we are interested in the following two probabilistic distributions:

$$u_i = \Pr[E \text{ occurs at the } i\text{th trial}], \quad i = 1, 2, \ldots,$$
$$f_i = \Pr[E \text{ occurs for the first time at the } i\text{th trial}], \quad i = 1, 2, \ldots.$$

Then, the mean waiting time between two successive occurrences of E is

$$\mu_E = \sum_i i f_i. \tag{B.44}$$

Theorem B.2. *If μ_E is finite, then $u_n \to \frac{1}{\mu_E}$ as $n \to \infty$.*

Define the indicator variable I_j for each j as

$$I_j = \begin{cases} 1 & \text{if } E \text{ occurs at the } j\text{th trial}, \\ 0 & \text{if it does not.} \end{cases}$$

Then, the sum

$$X_k = I_1 + I_2 + \ldots + I_k$$

denotes the number of occurrences of E in the first k trials. By the linearity property of the mean, we have

$$E(N_k) = E(I_1) + E(I_2) + \cdots + E(I_k) = u_1 + u_2 + \cdots + u_k.$$

If μ_E is finite, by Theorem B.2, we have

$$E(X_N) \approx \frac{N}{\mu_E}. \tag{B.45}$$

Given two arbitrary sequences

$$b_0, b_1, \ldots,$$
$$f_1, f_2, \ldots.$$

We define the third sequence by convolution equation

$$v_n = b_n + v_{n-1} f_1 + \cdots + v_0 f_n, \quad n = 0, 1, \ldots.$$

The following theorem provides conditions under which the sequence v_n defined through convolution equation converges as n goes to infinity.

Theorem B.3. *(Renewal Theorem) If* $\sum_i b_i = B < \infty$, $\sum_i f_i = 1$, *and* $\mu = \sum_i i f_i < +\infty$, *and the greatest common divisor of indices i such that $f_i \neq 0$ is 1, then*

$$v_n \longrightarrow \frac{B}{\mu}$$

as $n \longrightarrow \infty$.

Appendix C
Software Packages for Sequence Alignment

In this appendix, we compile a list of software packages for pairwise alignment, homology search, and multiple alignment.

Table C.1 Software packages for pairwise alignment.

Package Name	Website
ARIADNE [145]	http://www.well.ox.ac.uk/ariadne
band [43]	http://globin.bx.psu.edu/dist/band
BLASTZ [177]	http://www.bx.psu.edu/miller_lab/dist/blastz-2004-12-27.tar.gz
GAP3 [93]	http://deepc2.psi.iastate.edu/aat/align/align.html
MUMmer [119]	http://mummer.sourceforge.net
robust [40]	http://globin.bx.psu.edu/dist/robust
PipMaker [178]	http://bio.cse.psu.edu/pipmaker
SIM [92]	http://ca.expasy.org/tools/sim.html
SIM4 [73]	http://globin.bx.psu.edu/dist/sim4/sim4.tar.gz
SSEARCH [180]	http://fasta.bioch.virginia.edu/fasta_www2/fasta_list2.shtml

Table C.2 Software packages for homology search.

Package Name	Website
BLAST [7, 8]	http://www.ncbi.nlm.nih.gov/blast
BLAT [109]	http://genome.ucsc.edu/cgi-bin/hgBlat
FASTA [161]	http://fasta.bioch.virginia.edu/fasta_www2/fasta_list2.shtml
MEGABLAST [215]	http://www.ncbi.nlm.nih.gov/blast/megablast.shtml
PatternHunter [124, 131]	http://www.bioinformaticssolutions.com/products/ph
WU-BLAST	http://blast.wustl.edu

Table C.3 Software packages for multiple alignment.

Package Name	Website
ClustalW [120, 189]	http://www.ebi.ac.uk/Tools/clustalw2/index.html
DIALIGN [143]	http://dialign.gobics.de
Kalign [121]	http://msa.sbc.su.se
MAFFT [106]	http://align.bmr.kyushu-u.ac.jp/mafft/online/server
MAP2 [210]	http://deepc2.psi.iastate.edu/aat/map/map.html
MAVID [28]	http://baboon.math.berkeley.edu/mavid
MSA [84, 127]	http://www.psc.edu/general/software/packages/msa/manual
Multi-LAGAN [32]	http://lagan.stanford.edu/lagan_web/index.shtml
MultiPipMaker [178]	http://bio.cse.psu.edu/pipmaker
MUMMALS [163]	http://prodata.swmed.edu/mummals
MUSCLE [62, 63]	http://www.drive5.com/muscle
ProbCons [59]	http://probcons.stanford.edu
PROMALS [164]	http://prodata.swmed.edu/promals
PSAlign [187]	http://faculty.cs.tamu.edu/shsze/psalign
T-Coffee [155]	http://www.tcoffee.org
T-Lara [24]	https://www.mi.fu-berlin.de/w/LiSA/Lara
YAMA [41]	http://globin.bx.psu.edu/dist/yama

References

1. Altschul, S.F. (1989) Generalized affine gap costs for protein sequence alignment. *Proteins* **32**, 88-96.
2. Altschul, S.F. (1989) Gap costs for multiple sequence alignment. *J. Theor. Biol.* **138**, 297-309.
3. Altschul, S.F. (1991) Amino acid substitution matrices from an information theoretic perspective. *J Mol. Biol.* **219**, 555-65.
4. Altschul, S.F., Boguski, M.S., Gish, W., and Wootton, J.C. (1994) Issues in searching molecular sequence databases. *Nat. Genet.* **6**, 119-129.
5. Altschul, S.F., Bundschuh, R., Olsen, R., and Hwa, T. (2001) The estimation of statistical parameters for local alignment score distributions. *Nucleic Acids Res.* **29**, 351-361.
6. Altschul, S.F. and Gish, W. (1996) Local alignment statistics. *Methods Enzymol.* **266**, 460-480.
7. Altschul, S.F., Gish, W., Miller, W., Myers, E., and Lipman, D.J. (1990) Basic local alignment search tool. *J. Mol. Biol.* **215**, 403-410.
8. Altschul, S.F., Madden, T.L., Schäffer, A.A., Zhang, J., Zhang, Z., Miller, W., and Lipman, D.J. (1997) Gapped Blast and Psi-Blast: a new generation of protein database search programs. *Nucleic Acids Res.* **25**, 3389-3402.
9. Altschul, S.F., Wootton, J.C., Gertz, E.M., Agarwala, R., Morgulis, A., Schaffer, A.A., and Yu, Y.K. (2005) Protein database searches using compositionally adjusted substitution matrices. *FEBS J.* **272**, 5101-5109.
10. Arratia, R., Gordon, L., and Waterman, M.S. (1986) An extreme value theory for sequence matching. *Ann. Stat.* **14**, 971-983.
11. Arratia, R., Gordon, L., and Waterman, M.S. (1990) The Erdös-Renýi law in distribution, for coin tossing and sequence matching. *Ann. Stat.* **18**, 539-570.
12. Arratia, R. and Waterman, M.S. (1985) An Erdös-Renýi law with shifts. *Adv. Math.* **55**, 13-23.
13. Arratia, R. and Waterman, M.S. (1986) Critical phenomena in sequence matching. *Ann. Probab.* **13**, 1236-1249.
14. Arratia, R. and Waterman, M.S. (1989) The Erdös-Renýi strong law for pattern matching with a given proportion of mismatches. *Ann. Probab.* **17**, 1152-1169.
15. Arratia, R. and Waterman, M.S. (1994) A phase transition for the scores in matching random sequences allowing deletions. *Ann. Appl. Probab.* **4**, 200-225.
16. Arvestad, L. (2006), Efficient method for estimating amino acid replacement rates. *J. Mol. Evol.* **62**, 663-673.
17. Baase, S. and Gelder, A.V. (2000) *Computer Algorithms - Introduction to Design and Analysis.* Addison-Wesley Publishing Company, Reading, Massachusetts.
18. Bafna, V., Lawler, E.L., and Pevzner, P.A. (1997) Approximation algorithms for multiple sequence alignment. *Theor. Comput. Sci.* **182**, 233-244.

197

19. Bafna, V. and Pevzner, P.A. (1996) Genome rearrangements and sorting by reversals. *SIAM J. Comput.* **25**, 272-289.
20. Bahr, A., Thompson, J.D., Thierry, J.C., and Poch, O. (2001) BAliBASE (Benchmark Alignment dataBASE): enhancements for repeats. transmembrane sequences and circular permutations. *Nucleic Acids Res.* **29**, 323-326.
21. Bailey, T.L. and Gribskov, M. (2002) Estimating and evaluating the statistics of gapped local-alignment scores. *J. Comput. Biol.* **9**, 575-593.
22. Balakrishnan, N. and Koutras, M.V. (2002) *Runs and Scans with Applications.* John Wiley & Sons, New York.
23. Batzoglou, S. (2005) The many faces of sequence alignment. *Brief. Bioinform.* **6**, 6-22.
24. Bauer, M., Klau, G.W., and Reinert, K. (2007) Accurate multiple sequence-structure alignment of RNA sequences using combinatorial optimization. *BMC Bioinformatics* **8**, 271.
25. Bellman, R. (1957) *Dynamic Programming.* Princeton University Press, Princeton, New Jersey.
26. Benner, S.A., Cohen, M.A., and Gonnet, G.H. (1993) Empirical and structural models for insertions and deletions in the divergent evolution of proteinsm. *J. Mol. Biol.* **229**, 1065-1082.
27. Bentley, J. (1986) *Programming Pearls.* Addison-Wesley Publishing Company, Reading, Massachusetts.
28. Bray, N. and Pachter, L. (2004) MAVID: Constrained ancestral alignment of multiple sequences. *Genome Res.* **14**, 693-699.
29. Brejovà, B., Brown D., and Vinař, T. (2004) Optimal spaced seeds for homologous coding regions. *J. Bioinform. Comput. Biol.* **1**, 595-610.
30. Brejovà, B., Brown, D., and Vinař, T. (2005) Vector seeds: an extension to spaced seeds. *J. Comput. Sys. Sci.* **70**, 364-380.
31. Brown, D.G. (2005) Optimizing multiple seed for protein homology search. *IEEE/ACM Trans. Comput. Biol. and Bioinform.* **2**, 29-38.
32. Brudno, M., Do, C., Cooper, G., Kim, M.F., Davydov, E., Green, E.D., Sidow, A., and Batzoglou, S. (2003) LAGAN and Multi-LAGAN: efficient tools for large-scale multiple alignment of genomic DNA. *Genome Res.* **13**, 721-731.
33. Buhler, J. (2001) Efficient large-scale sequence comparison by locality-sensitive hashing. *Bioinformatics* **17**, 419-428.
34. Buhler, J., Keich, U., and Sun, Y. (2005) Designing seeds for similarity search in genomic DNA. *J. Comput. Sys. Sci.* **70**, 342-363.
35. Bundschuh, R. (2002) Rapid significance estimation in local sequence alignment with gaps. *J. Comput. Biol.* **9**, 243-260.
36. Burkhardt, S. and Kärkkäinen, J. (2003) Better filtering with gapped q-grams. *Fund. Inform.* **56**, 51-70.
37. Califano, A. and Rigoutsos, I. (1993) FLASH: A fast look-up algorithm for string homology. In *Proc. 1st Int. Conf. Intell. Sys. Mol. Biol.*, AAAI Press, pp. 56-64.
38. Carrilo, H. and Lipman, D. (1988) The multiple sequence alignment problem in biology. *SIAM J. Applied Math.* **48**, 1073-1082.
39. Chan, H.P. (2003) Upper bounds and importance sampling of *p*-values for DNA and protein sequence alignments. *Bernoulli* **9**, 183-199
40. Chao, K.-M., Hardison, R.C., and Miller, W. (1993) Locating well-conserved regions within a pairwise alignment. *Comput. Appl. Biosci.* **9**, 387-396.
41. Chao, K.-M., Hardison, R.C., and Miller, W. (1994) Recent developments in linear-space alignment methods: a survey. *J. Comput. Biol.* **1**, 271-291.
42. Chao, K.-M. and Miller, W. (1995) Linear-space algorithms that build local alignments from fragments. *Algorithmica* **13**, 106-134.
43. Chao, K.-M., Pearson, W.R., and Miller, W. (1992) Aligning two sequences within a specified diagonal band. *Comput. Appl. Biosci.* **8**, 481-487.
44. Chen, L. (1975) Poisson approximation for dependent trials. *Ann. Probab.* **3**, 534-545.
45. Chiaromonte, F., Yap, V.B., and Miller, W. (2002) Scoring pairwise genomic sequence alignments. In *Proc. Pac. Symp. Biocomput.*, 115-126.

46. Chiaromonte, F., Yang, S., Elnitski, L., Yap, V.B., Miller, W., and Hardison, R.C. (2001) Association between divergence and interspersed repeats in mammalian noncoding genomic DNA. *Proc. Nat'l. Acad. Sci. USA* **98**, 14503-8.
47. Choi, K.P. and Zhang, L.X. (2004) Sensitivity analysis and efficient method for identifying optimal spaced seeds. *J. Comput. Sys. Sci.* **68**, 22-40.
48. Choi, K.P., Zeng, F., and Zhang L.X. (2004) Good spaced seeds for homology search. *Bioinformatics* **20**, 1053-1059.
49. Chvtal V and Sankoff D. (1975) Longest common subsequence of two random sequences. *J. Appl. Probab.* **12**, 306-315.
50. Coles, S. (2001) *An Introduction to Statistical Modeling of Extreme Values.* Springer-Verlag, London, UK.
51. Collins, J.F., Coulson, A.F.W., and Lyall, A. (1998) The significance of protein sequence similarities. *Comput. Appl. Biosci.* **4**, 67-71.
52. Cormen, T.H., Leiserson, C.E., Rivest, R.L., and Stein, C. (2001) *Introduction to Algorithms.* The MIT Press, Cambridge, Massachusetts.
53. Csürös, M., and Ma, B. (2007) Rapid homology search with neighbor seeds. *Algorithmica* **48**, 187-202.
54. Darling, A., Treangen, T., Zhang, L.X., Kuiken, C., Messeguer, X., and Perna, N. (2006) Procrastination leads to efficient filtration for local multiple alignment. In *Proc. 6th Int. Workshop Algorithms Bioinform. Lecture Notes in Bioinform.*, vol. 4175, pp.126-137.
55. Dayhoff, M.O., Schwartz, R.M., and Orcutt, B.C. (1978). A model of evolutionary changes in proteins. In *Atlas of Protein Sequence and Structure* vol. 5, suppl 3 (ed. M.O. Dayhoff), 345-352, Nat'l Biomed. Res. Found, Washington.
56. Dembo, A., Karlin, S., and Zeitouni, O. (1994) Critical phenomena for sequence matching with scoring. *Ann. Probab.* **22**, 1993-2021.
57. Dembo, A., Karlin, S., and Zeitouni, O. (1994) Limit distribution of maximal non-aligned two sequence segmental score. *Ann. Probab.* **22**, 2022-2039.
58. Deonier, R.C., Tavaré, S., and Waterman, M.S. (2005), *Computational Genome Analysis.* Springer, New York.
59. Do, C.B., Mahabhashyam, M.S.P., Brudno, M., and Batzoglou, S. (2005) PROBCONS: Probabilistic consistency-based multiple sequence alignment. *Genome Res.* **15**, 330-340.
60. Dobzhansky, T. and Sturtevant, A.H. (1938) Inversions in the chromosomes of Drosophila pseudoobscura. *Genetics* **23**, 28-64.
61. Durbin, R., Eddy, S., Krogh, A., and Mitichison, G. (1998) *Biological Sequence Analysis: Probabilistic Models of Protein and Nucleic Acids.* Cambridge University Press, Cambridge, UK.
62. Edgar, R.C. (2004) MUSCLE: multiple sequence alignment with high accuracy and high throughput. *Nucleic Acids Res.* **32**, 1792-1797.
63. Edgar, R.C. (2004) MUSCLE: a multiple sequence alignment method with reduced time and space complexity. *BMC Bioinformatics* **5**, no. 113.
64. Ewens, W.J. and Grant, G.R. (2001) *Statistical Methods in Bioinformatics: An Introduction.* Springer-Verlag, New York.
65. Farach-Colton, M., Landau, G., Sahinalp, S.C., and Tsur, D. (2007) Optimal spaced seeds for faster approximate string matching. *J. Comput. Sys. Sci.* **73**, 1035-1044.
66. Fayyaz, A.M., Mercier, S., Ferré, Hassenforder, C. (2008) New approximate *P*-value of gapped local sequence alignments. *C. R. Acad. Sci. Paris Ser. I* **346**, 87-92.
67. Feller, W. (1966) *An introduction to Probability Theory and its Applications.* Vol. 2 (1st edition), John Wiley & Sons, New York.
68. Feller, W. (1968) *An introduction to Probability Theory and its Applications.* Vol. I (3rd edition), John Wiley & Sons, New York.
69. Feng, D.F. and Doolittle, R.F. (1987) Progressive sequence alignment as a prerequisite to correct phylogenetic trees. *J. Mol. Evol.* **25**, 351-360.
70. Feng, D.F. Johnson, M.S. and Doolittle, R.F. (1985) Aligning amino acid sequences: comparison of commonly used methods. *J. Mol. Evol.* **21**, 112-125.

71. Fitch, W.M. and Smith, T.F. (1983) Optimal sequence alignments. *Proc. Nat'l. Acad. Sci. USA* **80**, 1382-1386.

72. Flannick, J., and Batzoglou, S. (2005) Using multiple alignments to improve seeded local alignment algorithms. *Nucleic Acids Res.* **33**, 4563-4577.

73. Florea, L., Hartzell, G., Zhang, Z., Rubin, G.M., and Miller W. (1998) A computer program for aligning a cDNA sequence with a genomic DNA sequence. *Genome Res.* **8**, 967-974.

74. Gertz, E.M. (2005) BLAST scoring parameters. *Manuscript*(ftp://ftp.ncbi.nlm.nih.gov/blast/documents/developer/scoring.pdf).

75. Giegerich, R. and Kurtz, S. (1997) From Ukkonen to McCreight and Weiner: A unifying view of linear-time suffix tree construction. *Algorithmica* **19**, 331-353.

76. Gilbert, W. (1991) Towards a paradigm shift in biology. *Nature* **349**, 99.

77. Gonnet, G.H., Cohen, M.A., and Benner, S.A. (1992) Exhaustive matching of the entire protein sequence database. *Science* **256**, 1443-5.

78. Gotoh, O. (1982) An improved algorithm for matching biological sequences. *J. Mol. Biol.* **162**, 705-708.

79. Gotoh, O. (1989) Optimal sequence alignment allowing for long gaps. *Bull. Math. Biol.* **52**, 359-373.

80. Gotoh, O. (1996) Significant improvement in accuracy of multiple protein sequence alignments by iterative refinement as assessed by reference to structural alignments. *J. Mol. Biol.* **264**, 823-838.

81. Gribskov, M., Luthy, R., and Eisenberg, D. (1990) Profile analysis. In R.F. Doolittle (ed.) *Molecular Evolution: Computer Analysis of Protein and Nucleic Acid Sequences*, Methods in Enzymol., vol. 183, Academic Press, New York, pp. 146-159.

82. Grossman, S. and Yakir, B. (2004) Large deviations for global maxima of independent superadditive processes with negative drift and an application to optimal sequence alignment. *Bernoulli* **10**, 829-845.

83. Guibas, L.J. Odlyzko, A.M. (1981) String overlaps, pattern matching, and nontransitive games. *J. Combin. Theory* (series A) **30**, 183-208.

84. Gupta, S.K., Kececioglu, J., and Schaffer, A.A. (1995) Improving the practical space and time efficiency of the shortest-paths approach to sum-of-pairs multiple sequence alignment. *J. Comput. Biol.* **2**, 459-472.

85. Gusfield, D. (1997) *Algorithms on Strings, Trees, and Sequences*. Cambridge University Press, Cambridge, UK.

86. Hannenhalli, S. and Pevzner, P.A. (1999) Transforming cabbage into turnip (polynomial algorithm for sorting signed permutations by reversals). *J. Assoc. Comput. Mach.* **46**, 1-27.

87. Hardy, G., Littlewood, J.E., and Pólya, G. (1952) *Inequalities*, Cambridge University Press, Cambridge, UK.

88. Henikoff, S. and Henikoff, JG (1992) Amino acid substitution matrices from protein blocks. *Proc. Nat'l Acad. Sci. USA* **89**, 10915-10919.

89. Henikoff, S. and Henikoff, JG (1993) Performance evaluation of amino acid substitution matrices. *Proteins* **17**, 49-61.

90. Hirosawa, M., Totoki, Y., Hoshida, M., and Ishikawa, M. (1995) Comprehensive study on iterative algorithms of multiple sequence alignment. *Comput. Appl. Biosci.* **11**, 13-18.

91. Hirschberg, D.S. (1975) A linear space algorithm for computing maximal common subsequences. *Comm. Assoc. Comput. Mach.* **18**, 341-343.

92. Huang, X. and Miller, W. (1991) A time-efficient, linear-space local similarity algorithm. *Adv. Appl. Math.* **12**, 337-357.

93. Huang, X. and Chao, K.-M. (2003) A generalized global alignment algorithm. *Bioinformatics* **19**, 228-233.

94. Huffman, D.A. (1952) A method for the construction of minimum-redundancy codes. *Proc. IRE* **40**, 1098-1101.

95. Ilie, L., and Ilie, S. (2007) Multiple spaced seeds for homology search. *Bioinformatics* **23**, 2969-2977

96. Indyk, P. and Motwani, R. (1998) Approximate nearest neighbors: towards removing the curse of dimensionality. In *Proc. 30th Ann. ACM Symp. Theory Comput.*, 604-613.

97. Jones D.T., Taylor, W.R., and Thornton, J.M. (1992) The rapid generation of mutation data matrices from protein sequences. *Comput. Appl. Biosci.* **8**, 275-82.

98. Jones, N.C. and Pevzner, P.A. (2004) *Introduction to Bioinformatics Algorithms.* The MIT Press, Cambridge, Massachusetts.

99. Karlin, S. (2005) Statistical signals in bioinformatics. *Proc Nat'l Acad Sci USA.* **102**, 13355-13362.

100. Karlin, S. and Altschul, S.F. (1990) Methods for assessing the statistical significance of molecular sequence features by using general scoring schemes. *Proc. Nat'l. Acad. Sci. USA* **87**, 2264-2268.

101. Karlin, S. and Altschul, S.F. (1993) Applications and statistics fro multiple high-scoring segments in molecular sequences. *Proc. Nat'l Acad. Sci. USA* **90**, 5873-5877.

102. Karlin, S. and Dembo, A. (1992) Limit distribution of maximal segmental score among Markov-dependent partial sums. *Adv. Appl. Probab.* **24**, 113-140.

103. Karlin, S. and Ost, F. (1988) Maximal length of common words among random letter sequences. *Ann. Probab.* **16**, 535-563.

104. Karolchik, D., Kuhn, R.M., Baertsch, R., Barber, G.P., Clawson, H., Diekhans, M., Giardine, B., Harte, R.A., Hinrichs, A.S., Hsu, F., Miller, W., Pedersen, J.S., Pohl, A., Raney, B.J., Rhead, B., Rosenbloom, K.R., Smith, K.E., Stanke, M., Thakkapallayil, A., Trumbower, H., Wang, T., Zweig, A.S., Haussler, D., Kent, W.J. (2008) The UCSC genome browser database: 2008 update. *Nucleic Acids Res.* **36**, D773-779.

105. Karp, R.M., and Rabin, M.O. (1987) Efficient randomized pattern-matching algorithms. *IBM J. Res. Dev.* **31**, 249-260.

106. Katoh, K., Kuma, K., Toh, H., and Miyata, T. (2005) MAFFT version 5: improvement in accuracy of multiple sequence alignment, *Nucleic Acids Res.* **20**, 511-518.

107. Kececioglu, J. and Starrett, D. (2004) Aligning alignments exactly. In *Proc. RECOMB*, 85-96.

108. Keich, U., Li, M., Ma, B., and Tromp, J. (2004) On spaced seeds for similarity search. *Discrete Appl. Math.* **3**, 253-263.

109. Kent, W.J. (2002) BLAT: The BLAST-like alignment tool. *Genome Res.* **12**, 656-664.

110. Kent, W.J. and Zahler, A.M. (2000) Conservation, regulation, synteny, and introns in a large-scale C. briggsae-C. elegans Genomic Alignment. *Genome Res.* **10**, 1115-1125.

111. Kimura, M. (1983) *The Neutral Theory of Molecular Evolution*, Cambridge University Press, Cambridge, UK.

112. Kisman, D., Li, M., Ma, B., and Wang, L. (2005) tPatternHunter: gapped, fast and sensitive translated homology search. *Bioinformatics* **21**, 542-544.

113. Knuth, D.E. (1973) *The art of computer programming.* Vol. 1, Addison-Wesley Publishing Company, Reading, Massachusetts.

114. Knuth, D.E. (1973) *The art of computer programming.* Vol. 3, Addison-Wesley Publishing Company, Reading, Massachusetts.

115. Kong, Y. (2007), Generalized correlation functions and their applications in selection of optimal multiple spaced seeds for homology search. *J. Comput. Biol.* **14**, 238-254.

116. Korf, I., Yandell, M., and Bedell, J. (2003) *BLAST.* O'reilly, USA.

117. Kschischo, M., Lässig, M., and Yu, Y.-K. (2005) Towards an accurate statistics of gapped alignment. *Bull. Math. Biol.* **67**, 169-191.

118. Kucherov G., Noè, L. and M. Roytberg M. (2005) Multiseed lossless filtration. *IEEE/ACM Trans. Comput. Biol. Bioinform.* **2**, 51-61.

119. Kurtz, S., Phillippy, A., Delcher, A.L., Smoot, M., Shumway, M., Antonescu, C., and Salzberg S.L. (2004) Versatile and open software for comparing large genomes. *Genome Biology* **5**, R12.

120. Larkin, M.A., Blackshields, G., Brown, N.P., Chenna, R., McGettigan, P.A., McWilliam, H., Valentin, F., Wallace, I.M., Wilm, A., Lopez, R., Thompson, J.D., Gibson, T.J., and Higgins, D.G. (2007) ClustalW and ClustalX version 2. *Bioinformatics* **23**, 2947-2948.

121. Lassmann, T. and Sonnhammer, L.L. (2005) Kalign - an accurate and fast multiple sequence alignment algorithm. *BMC Bioinformatics* **6**, 298.

122. Letunic, I., Copley, R.R., Pils, B., Pinkert, S., Schultz, J., Bork, P. (2006) SMART 5: domains in the context of genomes and networks. *Nucleic Acids Res.* **34**, D257-260.

123. Li, M., Ma, B., Kisman, D., and Tromp, J. (2004) PatternHunter II: Highly sensitive and fast homology search. *J. Bioinform. Comput. Biol.* **2**, 417-439.

124. Li, M., Ma, B. Kisman, D. and Tromp, J. (2004) PatternHunter II: Highly Sensitive and Fast Homology Search. *J. Bioinform Comput. Biol.* **2** (3),417-439.

125. Li, M., Ma, B. and Zhang, L.X. (2006) Superiority and complexity of spaced seeds. In *Proc. 17th SIAM-ACM Symp. Discrete Algorithms.* 444-453.

126. Li, W.H., Wu, C.I., and Luo, C.C. (1985) A new method for estimating synonymous and nonsynonymous rates of nucleotide substitution considering the relative likelihood of nucleotide and codon changes. *Mol. Biol. Evol.* **2**, 150-174.

127. Lipman, D.J., Altschul, S.F., and Kececioglu, J.D. (1989) A tool for multiple sequence alignment. *Proc. Nat'l. Acad. Sci. USA* **86**, 4412-4415.

128. Lipman, D.J. and Pearson, W.R. (1985) Rapid and sensitive protein similarity searches. *Science* **227**, 1435-1441.

129. Lothaire, M. (2005) *Applied Combinatorics on Words.* Cambridge University Press, Cambridge, UK.

130. Ma, B. and Li, M. (2007) On the complexity of spaced seeds. *J. Comput. Sys. Sci.* **73**, 1024-1034.

131. Ma, B., Tromp, J., and Li, M. (2002) PatternHunter-faster and more sensitive homology search. *Bioinformatics* **18**, 440-445.

132. Ma, B., Wang, Z., and Zhang, K. (2003) Alignment between two multiple alignments. *Proc. Combin. Pattern Matching*, 254-265.

133. Manber, U. (1989) *Introduction to Algorithms.* Addison-Wesley Publishing Company, Reading, Massachusetts.

134. Manber, U. and Myers, E. (1991) Suffix arrays: a new method for on-line string searches. *SIAM J. Comput.* **22**, 935-948.

135. McCreight, E.M. (1976) A space-economical suffix tree construction algorithm. *J. Assoc. Comput. Mach.* **23**, 262-272.

136. McLachlan, A.D. (1971) Tests for comparing related amino-acid sequences. Cytochrome c and cytochrome c551. *J. Mol. Biol.* **61**, 409-424.

137. Metzler, D. (2006) Robust E-values for gapped local alignment. *J. Comput. Biol.* **13**, 882-896.

138. Miller, W. (2001) Comparison of genomic DNA sequences: solved and unsolved problems. *Bioinformatics* **17**, 391-397.

139. Miller, W. and Myers, E. (1988) Sequence comparison with concave weighting functions. *Bull. Math. Biol.* **50**, 97-120.

140. Miller, W., Rosenbloom, K., Hardison, R.C., Hou, M., Taylor, J., Raney, B., Burhans, R., King, D.C., Baertsch, R., Blankenberg, D., Kosakovsky Pond, S.L., Nekrutenko, A., Giardine, B., Harris, R.S., Tyekucheva, S., Diekhans, M., Pringle, T.H., Murphy, W.J., Lesk, A., Weinstock, G.M., Lindblad-Toh, K., Gibbs, R.A., Lander, E.S., Siepel, A., Haussler, D., and Kent, W.J. (2007) 28-way vertebrate alignment and conservation track in the UCSC Genome Browser. *Genome Res.* **17**,1797-1808.

141. Mitrophanov, A.Y. and Borodovsky, M. (2006) Statistical significance in biological sequence analysis. *Brief. Bioinform.* **7**, 2-24.

142. Mohana Rao, J.K. (1987) New scoring matrix for amino acid residue exchanges based on residue characteristic physical parameters. *Int. J. Peptide Protein Res.* **29**, 276-281.

143. Morgenstern, B., French, K., Dress, A., and Werner, T. (1998) DIALIGN: Finding local similarities by multiple sequence alignment. *Bioinformatics* **14**, 290-294.

144. Mott, R. (1992) Maximum-likelihood estimation of the statistical distribution of Smith-Waterman local sequence similarity scores. *Bull. Math. Biol.* **54**, 59-75.

145. Mott, R. (2000) Accurate formula for P-values for gapped local sequence and profile alignment. *J. Mol. Biol.* **276**, 71-84.

146. Mott, R and Tribe, R. (1999) Approximate statistics of gapped alignments. *J. Comput. Biol.* **6**, 91-112.

147. Müller, T. and Vingron, M. (2000) Modeling amino acid replacement. *J. Comput. Biol.* **7**, 761-776.
148. Müller, T., Spang, R., and Vingron, M. (2002) Estimating amino acid substitution models: A comparison of Dayoff's estimator, the resolvent approach and a maximum likelihood method. *Mol. Biol. Evol.* **19**, 8-13.
149. Myers, G. (1999) Whole-Genome DNA sequencing. *Comput. Sci. Eng.* **1**, 33-43.
150. Myers, E. and Miller, W. (1988) Optimal alignments in linear space. *Comput. Appl. Biosci.* **4**, 11-17.
151. Needleman, S.B. and Wunsch, C.D. (1970) A general method applicable to the search for similarities in the amino acid sequences of two proteins. *J. Mol. Biol.* **48**, 443-453.
152. Nicodème, P., Salvy, B., and Flajolet, P. (1999) Motif Statistics. In *Lecture Notes in Comput. Sci.*, vol. 1643, 194-211, New York.
153. Noè, L., and Kucherov, G. (2004) Improved hit criteria for DNA local alignment. *BMC Bioinformatics* **5**, no.159.
154. Notredame, C. (2007) Recent evolutions of multiple sequence alignment algorithms. *PLoS Comput. Biol.* **3**, e123.
155. Notredame, C., Higgins, D., and Heringa, J. (2000) T-Coffee: A novel method for multiple sequence alignments. *J. Mol. Biol.* **302**, 205-217.
156. Overington, J., Donnelly, D., Johnson, M.S., Sali, A., and Blundell, T.L. (1992) Environment-specific amino acid substitution tables: tertiary templates and prediction of protein folds. *Protein Sci.* **1**, 216-26.
157. Park, Y. and Spouge, J.L. (2002) The correction error and finite-size correction in an ungapped sequence alignment. *Bioinformatics* **18**, 1236-1242.
158. Pascarella, S. and Argos, P. (1992) Analysis of insertions/deletions in protein structures, *J. Mol. Biol.* **224**, 461-471.
159. Pearson, W.R. (1995) Comparison of methods for searching protein sequence databases. *Protein Science* **4**, 1145-1160.
160. Pearson, W.R. (1998) Empirical statistical estimates for sequence similarity searches. *J. Mol. Biol.* **276**, 71-84.
161. Pearson, W.R. and Lipman, D. (1988) Improved tools for biological sequence comparison. *Proc. Nat'l. Acad. Sci USA* **85**, 2444-2448.
162. Pearson, W.R. and Wood, T.C. (2003) Statistical Significance in Biological Sequence Comparison. In *Handbook of Statistical Genetics* (edited by D.J. Balding, M. Bishop and C. Cannings), 2nd Edition. John Wiley & Sons, West Sussex, UK.
163. Pei, J. and Grishin, N.V. (2006) MUMMALS: multiple sequence alignment improved by using hidden Markov models with local structural information. *Nucleic Acids Res.* **34**, 4364-4374.
164. Pei, J. and Grishin, N.V. (2007) PROMALS: towards accurate multiple sequence alignments of distantly related proteins. *Bioinformatics* **23**, 802-808.
165. Pevzner, P.A. (2000) *Computational Molecular Biology: An Algorithmic Approach*. The MIT Press, Cambridge, Massachusetts.
166. Pevzner, P.A. and Tesler, G. (2003) Genome rearrangements in mammalian evolution: lessons from human and mouse genomes. *Genome Res.* **13**, 37-45.
167. Pevzner, P.A. and Waterman, M.S. (1995) Multiple filtration and approximate pattern matching. *Algorithmica* **13**, 135-154.
168. Preparata, F.P., Zhang, L.X., and Choi, K.P. (2005) Quick, practical selection of effective seeds for homology search. *J. Comput. Biol.* **12**, 1137-1152.
169. Reich, J.G., Drabsch, H., and Däumler, A. (1984) On the statistical assessment of similarities in DNA sequences. *Nucleic Acids Res.* **12**, 5529-5543.
170. Rényi, A. (1970) *Foundations of Probability*. Holden-Day, San Francisco.
171. Reinert, G., Schbath, S., and Waterman, M.S. (2000), Probabilistic and statistical properties of words: An overview. *J. Comput. Biol.* **7**, 1 - 46.
172. Risler, J.L., Delorme, M.O., Delacroix, H., and Henaut, A. (1988) Amino Acid substitutions in structurally related proteins. A pattern recognition approach. Determination of a new and efficient scoring matrix. *J. Mol. Biol.* **204**, 1019-1029.

173. Robinson, A.B. and Robinson, L.R. (1991) Distribution of glutamine and asparagine residues and their near neighbors in peptides and proteins. *Proc. Nat'l. Acad. Sci USA* **88**, 8880-8884.

174. Sankoff, D. (2000) The early introduction of dynamic programming into computational biology. *Bioinformatics* **16**, 41-47.

175. Sankoff, D. and Kruskal, J.B. (eds.) (1983). *Time Warps, String Edits, and Macromolecules: the Theory and Practice of Sequence Comparisons.* Addison-Wesley, Reading, Massachusetts.

176. Schwager, S.J. (1983) Run probabilities in sequences of Markov-dependent trials, *J. Amer. Stat. Assoc.* **78**, 168-175.

177. Schwartz, S., Kent, W.J., Smit, A., Zhang, Z., Baertsch, R., Hardison, R.C., Haussler, D., and Miller, W. (2003) Human-mouse alignment with BLASTZ. *Genome Res.* **13**, 103-107.

178. Schwartz, S., Zhang, Z., Frazer, K.A., Smit, A., Riemer, C., Bouck, J., Gibbs, R., Hardison, R.C., and Miller, W. (2000) PipMaker - a web server for aligning two genomic DNA sequences. *Genome Res.* **10**, 577-86.

179. Siegmund, D. and Yakir, B. (2000) Approximate *p*-value for local sequence alignments. *Ann. Stat.* **28**, 657-680.

180. Smith, T.F. and Waterman, M.S., (1981) Identification of common molecular subsequences. *J. Mol. Biol.* **147**, 195-197.

181. Smith, T.F., Waterman, M.S., and Burks, C. (1985) The statistical distribution of nucleic acid similarities. *Nucleic Acids Res.* **13**, 645-656.

182. Spang, R. and Vingron, M. (1998) Statistics of large-scale sequences searching. *Bioinformatics* **14**, 279-284.

183. Spitzer, F. (1960) A tauberian theorem and its probability interpretation. *Trans. Amer Math Soc.* **94**, 150-169.

184. States, D.J., Gish, W., and Altschul, S.F. (1991), Imporved sensitivity of nucleic acid databases searches using application-specific scoring matrices. *Methods* **3**, 61-70.

185. Sun, Y. and Buhler, J. Designing multiple simultaneous seeds for DNA similarity search. *J. Comput. Biol.* **12**, 847-861.

186. Sun, Y., and Buhler, J. (2006) Choosing the best heuristic for seeded alignment of DNA sequences. *BMC Bioinformatics* **7**, no. 133.

187. Sze, S.-H., Lu, Y., and Yang, Q. (2006) A polynomial time solvable formulation of multiple sequence alignment. *J. Comput. Biol.* **13**, 309-319.

188. Taylor, W.R. (1986) The classification of amino acid conservation. *J. Theor. Biol.* **119**, 205-218.

189. Thompson, J.D., Higgins, D.G., and Gibson, T.J. (1994) CLUSTAL W: improving the sensitivity of progressive multiple sequence alignment through sequence weighting, position-specific gap penalties and weight matrix choice. *Nucleic Acids Res.* **22**, 4673-4680.

190. Thompson, J.D., Plewniak, F., and Poch, O. (1999) A comprehensive comparison of multiple sequence alignment programs. *Nucleic Acids Res.* **27**, 2682-2690.

191. Ukkonen, E. (1995) On-line construction of suffix trees. *Algorithmica* **14**, 249-260.

192. Vingron, M. and Argos, P. (1990) Determination of reliable regions in protein sequence alignment. *Protein Eng.* **3**, 565-569.

193. Vingron, M. and Waterman, M.S. (1994) Sequence alignment and penalty choice: Review of concepts, case studies and implications. *J. Mol. Biol.* **235**, 1-12.

194. Wagner, R.A. and Fischer, M.J. (1974) The string-to-string correction problem. *J. Assoc. Comput. Mach.* **21**, 168-173.

195. Wang, L. and Jiang, T. (1994) On the complexity of multiple sequence alignment. *J. Comput. Biol.* **1**, 337-348.

196. Waterman, M.S. (1984) Efficient sequence alignment algorithms. *J. Theor. Biol.* **108**, 333-337.

197. Waterman, M.S. (1995) *Introduction to Computational Biology.* Chapman and Hall, New York.

198. Waterman, M. S. and Eggert, M. (1987) A new algorithm for best subsequence alignments with application to tRNA-rRNA comparisons. *Mol. Biol.* **197**, 723-728.

199. Waterman, W.S., Gordon, L., and Arratia, R. (1987) Phase transition in sequence matches and nucleic acid structure. *Proc. Nat'l. Acad. Sci. USA* **84**, 1239-1243.
200. Waterman, M.S. and Vingron, M. (1994) Rapid and accurate estimate of statistical significance for sequence database searches. *Proc. Nat'l. Acad. Sci. USA* **91**, 4625-4628.
201. Webber, C. and Barton, G.J. (2001) Estimation of *P*-value for global alignments of protein sequences. *Bioinformatics* **17**, 1158-1167.
202. Weiner, P. (1973) Linear pattern matching algorithms. In *Proc. the 14th IEEE Annu. Symp. on Switching and Automata Theory*, pp. 1-11.
203. Wilbur, W.J. (1985), On the PAM matrix model of protein evolution. *Mol. Biol. Evol.* **2**, 434-447.
204. Wilbur, W. and Lipman, D. (1983) Rapid similarity searches of nucleic acid and protein data banks. *Proc. Nat. Acad. Sci. USA* **80**, 726-730.
205. Wilbur, W. and Lipman, D. (1984) The context dependent comparison of biological sequences. *SIAM J. Appl. Math.* **44**, 557-567.
206. Xu, J., Brown, D. Li, M., and Ma, M. (2006) Optimizing multiple spaced seeds for homology search. *J. Comput. Biol.* **13**, 1355-1368.
207. Yang, I.-H., Wang, S.-H., Chen, Y.-H., Huang, P.-H., Ye, L., Huang, X. and Chao, K.-M. (2004) Efficient methods for generating optimal single and multiple spaced seeds. In *Proc. IEEE 4th Symp. on Bioinform. and Bioeng.*, pp. 411-418.
208. Yang J.L., and Zhang, L.X. (2008) Run probabilities of seed-like patterns and identifying good transition seeds. *J. Comput. Biol.* (in press).
209. Yap, V.B. and Speed, T. (2005), Estimating substitution matrices. In *Statistical Methods in Molecular Evolution* (ed. R. Nielsen), Springer.
210. Ye, L. and Huang, X. (2005) MAP2: Multiple alignment of syntenic genomic sequences. *Nucleic Acids Res.* **33**, 162-170.
211. Yu, Y.K., Wootton, J.C., and Altschul, S.F. (2003) The compositional adjustment of amino acid substitution matrices. *Proc. Nat'l. Acad. Sci. USA.* **100**, 15688-93.
212. Yu, Y.K. and Altschul, S.F. (2005) The construction of amino acid substitution matrices for the comparison of proteins with non-standard compositions. *Bioinformatics* **21**, 902-11.
213. Zachariah, M.A., Crooks, G.E., Holbrook, S.R., Brenner, S.E. (2005) A generalized affine gap model significantly improves protein sequence alignment accuracy. *Proteins* **58**, 329-38.
214. Zhang, L.X. (2007) Superiority of spaced seeds for homology search. *IEEE/ACM Trans. Comput. Biol. Bioinform.* **4**, 496-505.
215. Zhang, Z, Schwartz, S, Wagner, L, and Miller, W. (2000) A greedy algorithm for aligning DNA sequences. *J. Comput. Biol.* **7**, 203-214.
216. Zhou L. and Florea, L. (2007) Designing sensitive and specific spaced seeds for cross-species mRNA-to-genome alignment. *J. Comput. Biol.* **14**, 113-130,
217. Zuker, M. and Somorjal, R.L. (1989) The alignment of protein structures in three dimensions. *Bull. Math. Biol.* **50**, 97-120.

Index